Slightly Higher In Canada

by **Jeff Pardo**

with a Foreword by Barry Schiff

INFINITY PUBLISHING

ISBN 978-0-7414-7524-4 Paperback
ISBN 978-0-7414-7525-1 eBook

Printed in the United States of America

Published June 2012

INFINITY PUBLISHING
1094 New DeHaven Street, Suite 100
West Conshohocken, PA 19428-2713
Toll-free (877) BUY BOOK
Local Phone (610) 941-9999
Fax (610) 941-9959
Info@buybooksontheweb.com
www.buybooksontheweb.com

Table of Contents

Foreword iii

Preface v

Chapter 1: History Lessons 1

Chapter 2: Greater Perspectives 62

Chapter 3: Airmen, Aircraft, Airports, and Airspace 69

Chapter 4: Airmanship and Aerodynamics 109

Chapter 5: Strange Weather 138

Chapter 6: Aerospace 168

Chapter 7: Avionics and Navigation 175

Chapter 8: Helicopters 193

Chapter 9: Accomplishments of Questionable Merit 200

Chapter 10: Curiosities, Factoids, and Helpful Hints 213

Chapter 11: Regulations In Question 238

Chapter 12: Things That Make You Think 257

Chapter 13: Other Things You'd *Never* Think About 282

About the Author 303

Foreword

When I took my first flying lesson in 1952 (at the tender and innocent age of 14), I began to discover a growing and insatiable desire for aviation knowledge. I don't mean that I just wanted to learn the facts necessary to become a safe and proficient pilot. I also had a burning need to learn aviation trivia and minutiae, the origins of colorful expressions, the logic behind rules of thumb, and the stories behind the plethora of wives tales.

For example, why is such a seemingly joyous word as *mayday* used by pilots to announce a dire emergency? (It comes from the French, *m'aidez*, which means *help me.*) Why is the control stick in the cockpit of an airplane called a *joystick*? (The word evolved in aviation long before the advent of the computer.) Why is a speed of one nautical mile per hour referred to as a knot? (These pages will unlock the mystery behind that and hundreds of other questions.)

It was as though my mind was a huge sponge. I couldn't get enough.

One of my first flight instructors, Paul Bell, used to point me out to others by saying, "There's Barry Schiff. He knows more [stuff] about flying that no one cares about than anyone else I know." This made me feel that there was something strange about my unbridled curiosity. Was I really so unique in this regard? I eventually learned that most pilots are indeed interested in such [stuff].

I discovered also and with some relief that I was not the only one with such a compulsive need to uncover aviation tidbits and acorns, which is a convenient and appropriate way to introduce you to someone similarly afflicted, Jeff Pardo, the author of the incredibly fascinating volume that you now hold before you. He, too, is one of those rare fountains of aviation knowledge and lore. He brings to light in this book an incredible amount of fascinating information that otherwise might fade away on the dusty, yellowing pages of early aviation books or the crumbling pages of old magazines.

The variety and breadth of information presented here is both entertaining and educational. Beyond that, Pardo has unearthed information that can make you a better and safer pilot.

I leave you now in the capable company of my friend and educator, Jeff Pardo. Have fun!

Barry Schiff
Captain, TWA (retired)
Los Angeles, California

Preface

Aviation can be an all-consuming affliction, and the quest for greater knowledge is a necessary precursor to acquiring the information and skills that make us all better pilots, or even peripherally, simply one of aviation's educated observers. At the outset, it might seem that a collection of ostensibly inconsequential details might hardly be considered as worthy of broadening our understanding and view of the world around us. I might rhetorically reply that, unlike many other pursuits, when it comes to the affairs of flight it is what you *don't* know that can hurt you. After all, as the saying goes, the Devil's in the details. And somewhat paradoxically, it often requires a broader perception of the world around us in order to appreciate the significance of those same details. However, although many of us enjoy the higher level of awareness that flight affords, we must nonetheless realize that not everyone shares it.

I am, by both outlook and disposition, one who questions what I see around me. Having the additional dubious blessing of being a lateral thinker, I revel in every unclaimed fact, unintended tangent, and incidental digression—knowledge for the sake of knowledge. While admirable in a species, its value to the individual might appear to rise no higher than fodder for social gatherings, or games within that long-lost and mysterious province of childhood. But by my own conjecture, I might counter with Nietzsche's admonition that the aim of all adults should be to rediscover in their work the seriousness of children at play. Having come under its uncanny spell, I invite you to join the game.

So, you have just become the beneficiary of a pedagogical menagerie of minutiae. Even if you live among the brotherhood of pilots, you may find some of these disjointed details to be profoundly useless. However, I have absolutely no doubt that in one way or another, more than a few will prove themselves instructive, useful, and at times amusing introductions to things that you might have had no idea you'd enjoy—or that would even need—knowing.

Chapter 1: History Lessons

We start off on our journey through some of the dustier corners of aviation history with this point of encouragement: History can be dull, but aviation history is almost never boring, especially when you look at some of the unusual exploits of our more forward-thinking forebears, as well as some of those whose thinking was better described (at least in hindsight) as wayward, rather than forward. To keep things fair for those with halfway decent peripheral vision, the answers are given at the end of each chapter, rather than immediately beneath each question (where one might unintentionally see the correct response).

Subject: Learning ... The Hard Way

Question 1: For fighter aircraft during WWII, what aspects of tracer rounds may have caused difficulty for aircrews?

A) Tracers were placed in every fifth round to aid in aiming. However, they had different ballistics, so at long range if your tracers were hitting the target, 80% of your rounds were missing.
B) Tracers instantly told your enemy that he was under fire, and from which direction.
C) The practice of loading a string of tracers at the end of the ammo belt told you that you were out of ammo. Unfortunately, it didn't take enemy aircraft crews long to figure this out for themselves.
D) All of the above

Subject: Famous First Words:

Question 2: Most people know about the (often misquoted) famous first words uttered by mankind on any heavenly body besides earth. They were spoken by Neil Armstrong, the first man to walk on the moon. They were of course "That's one small step for man; one giant leap for mankind". Right? Wrong! What WERE the REAL "first words"?

A) "That's one small step for a man, one giant leap for mankind." A static burst erased the "a". To this day, Neil Armstrong insists that he said it correctly.
B) Answer (A) is correct, with a qualifier. These really were the first words said outside the lunar module. However, the first words uttered in any communication with earth REALLY were: "Houston, Tranquility Base here. The Eagle has landed."
C) Answer (B) is more correct--sort of--but Neil doesn't have anything on his fellow astronaut, Buzz Aldrin, who actually was reading off a checklist (those checklists! See? They're everywhere!) At the time, there was concern that their ship might be swallowed up by a bottomless sea of

lunar dust, accumulated over eons of time. No, the first words were "Contact light. Okay. Engine stop."

D) This one isn't in the official archives. When Neil was a boy, he heard his next-door neighbor, Mrs. Mazur, chewing out her husband: "You want oral sex? I'll give you oral sex--when the boy next door walks on the moon!" In his own unique and wry humor, Neil actually did say "Good luck, Mr. Mazur!" These were really his first words.

Subject: Famous First flights

Question 3: In 1927, Charles Lindbergh captured the world's imagination, as well as the Orteig prize of $25,000---which is about three-quarters of a million in today's dollars---when he flew solo non-stop across the Atlantic Ocean. But this wasn't the first transatlantic air crossing. Who really <u>was</u> first?

A) Two British pilots, John Alcock and Arthur Brown, actually made the first non-stop crossing of the Atlantic by air in a twin-engine Vikers Vimy biplane in 1919 (though it wasn't from New York to Paris).
B) A British dirigible, actually. It made the flight from England to New Jersey (the "other way") in 1919. Then, just as matter-of-factly, it flew back.
C) (A) and (B).
D) Six men died trying to win the prize money. (The prize stipulated that it be a solo, non-stop flight from New York to Paris.) One of them actually did complete the flight--almost. He was lost just off Western Ireland when his plane ran out of fuel a half-mile from shore. He drowned in the pounding surf.

Subject: Spin History:

Question 4: Who was the first person to survive a spin?

A) Wilbur Wright, 1912
B) Lt. Wilfred Parke, Royal Navy, 1912
C) Frederick A. Lindemann, 1914
D) Waldo Pepper, 1923

Subject: Zulu, Shmulu

Question 5: Why do we call Coordinated Universal Time (which was formerly known as Greenwich Mean Time) "Zulu" time?

A) Because back in the days of British colonial hegemony, Bantu slaves from South Africa were used as timekeepers on the estates of landholders, and the slang became entrenched in the common language as meaning "British time". When the standard time zones were adopted late in the 19th century, the term "Zulu" stuck.

B) The National Institute of Standards and Technology divided the world (from the US perspective) into 24 time zones, each with its central meridian corresponding to Mean Solar Time. Each time zone was assigned a letter (excluding the letters J and K). The Greenwich Observatory just happened to be in the last one.

C) Choice (A) is ridiculous fiction; choice (B) is actually partially correct. But in 1883, when Krakatoa erupted (OK, blew up), the International Meridian Conference was forced into a hurried meeting (one year later--hah!) and decided that another major global conflagration might throw timekeeping and trade routes out of whack, so they had better establish internationally agreed upon time zones, just in case the skies went black all over again.

D) Choices (B) and (C) actually both have some truth to them, however preposterous that may sound. With the coming of faster and faster transport systems, and more advanced astronomical techniques, the International Meridian Conference (there really was one) in Washington DC in November of 1884 allocated 24 equatorially perpendicular wedges, each of one hour angle (15 degrees of longitude) in width. Letters were assigned to each, and the Greenwich observatory in England was taken as the reference time zone, and given the last letter, Z. The "Zulu" comes simply from the commonly accepted phonetic alphabet. Time zones are often altered markedly to conform somewhat to state borders. The East coast of the US is actually on "Romeo" time, Central time is "Sierra", and the mountain states and the West coast are Tango and Uniform. (Nobody uses them though.)

Subject: History's Worst Aircraft

Question 6: Which aircraft would likely 'take the cake' for having been the unforgivably worst machine ever to take to the skies (or try doing so)?

A) Completed in July 1903 and the brainchild of Samuel Pierpont Langley, world-renowned scientist and Secretary of the Smithsonian, the Langley Aerodrome was launched by catapult from a houseboat on the Potomac River in Washington DC. On two separate occasions, in October and again in December 1903 (nine days before the Wright Brothers flew), the Aerodrome roared off the catapult, and then, as you probably know, dropped unceremoniously straight down into the river like a ton of bricks.

B) Back in the days when we still weren't sure what airplanes were supposed to look like, Horatio Phillips designed a series of genteel Victorian steam-powered “multiplane” craft with dozens, sometimes hundreds, of stacked “venetian blind” wings, and which used an elliptical racetrack as a runway. More often than not, they collapsed in a madly flapping heap.

C) Somewhat apropos for the holiday season, a dapper gentleman from North Carolina named William Christmas, who was a redefinition of the word charlatan. Mr. Christmas managed to convince several wealthy -- albeit gullible -- people (and some simply desperate people) during the depths of World War One that first, he had credentials (which he didn't), and second, that his creation which he called 'the Christmas Bullet', could fly. Made of steel and plywood, they had an un-braced hardwood 'high wing' design, and two test flights, both of which killed the test pilots. They actually did fly, but moments after becoming airborne the wings would twist, flap, and then peel away from the heavy fuselage, which then plunged earthward. 'Doctor' Christmas died quietly in 1960 at age 94.

D) Even worse, the Japanese MXY7 'Ohka' (cherry blossom) was a rocket-propelled hand-flown Kamikaze torpedo loaded with high explosives, which was used in WWII. Most unforgivably of all, it was of course designed to kill its pilot. (Of course, so were all the other aircraft pressed into this divine calling.)

Subject: The First Birdman

Question 7: Who was the first person to fly in a heavier than air machine?

A) Orville Wright; December 17, 1903.
B) Clement Ader, in the Eole, October 9, 1890.
C) Felix du Temple, 1874
D) Sir George Cayley*, 1849

Subject: Fire When Ready, Gridley!

Question 8: The earliest use of armament aboard aircraft and the use of aircraft as military weapons occurred

A) on an experimental basis before World War I
B) on April 1st, 1915, when Jean Navarre of France scored his first victory with a rifle
C) on October 5, 1914, when a French gunner named Quenault shot down an enemy airplane with a machine gun mounted in the front of the engine nacelle
D) on August 14, 1914, when the French sent two Voisins to attack Zeppelin sheds at Metz-Frascaty

Subject: The Great E-Ticket Ride

Question 9: The world's first airliner was

A) the Benoist Type XIV "Air Boat", 1914

B) a Junkers J13, used by the airline Deutsche Luft Hansa (which became guess what) in 1919
C) Reinhold Platz's F-II, which entered service in 1920
D) the Fokker F.VII, used by Pan American Airways, and which began service between Key West and Havana in January 1928.

Subject: Divine Metamorphosis

Question 10: What was the Aichi M6A1 Seiran?

A) It would have been an extremely large (half of a Kilometer long, in fact) hollow dirigible containing a literal "runway in the sky" along its truss-supported, lighted, and flat inner surface. It was envisioned (but never built) as an alternative to an aircraft carrier by the Japanese during the last days of WW II.
B) It was Japan's imitation of the V2, intended to be launched from submarines near our shores. It was never perfected, and in fact sank two of their own ships.
C) A submarine launched strike aircraft built in small numbers by the Japanese near the end of WW II. Its wings pitched forward 90 degrees about the spar root, then folded back against the fuselage to fit in tubes less than 12 feet wide. They were capable of close to 300 mph, had over a 700 mile range, and could carry either a bomb or torpedo weighing close to 1800 pounds.
D) A concealed weapon issued to the Japanese, intended to be worn under a soldier's hat. Forseeing surrender in large numbers near the end of WWII, the idea was that as large groups of soldiers were marched out, hands clasped on top of their heads, they would stun their captors in one last-ditch mass attack by basically hitting themselves on the head and discharging what was esentially a head-mounted gun.

Subject: To Fly!

Question 11: Who was really the first American to fly?

A) Orville Wright, on December 17, 1903
B) George Cayley, 1849
C) Jean Blanchard, 1793
D) Edward Warren, 1784

Subject: Mach One Without Help From Gravity

Question 12: What was the first aircraft able to break the sound barrier while in level flight?

A) the F-100 Super Sabre

B) the F-101 Voodoo
C) the F-103
D) the F-104 Starfighter

Subject: Secret Societies

Question 13: Who and what are the Quiet Birdmen?

A) a branch of sailplane enthusiasts
B) a colloquial name for a local chapter of deaf pilots
C) The Ancient and Secret Order of Quiet Birdmen, a fraternal organization begun in 1919. Charles Lindbergh was a member.
D) a posthumous title for deceased EAA members

Subject: Markers In the Mists of Time

Question 14: True or false: We know exactly where the Wright Brothers' first flight began.

Subject: Buying the Farm

Question 15: Where did the expression "buying the farm" (in reference to fatal accidents) originate? (More than one may apply.)

A) The term "buying the farm" was around during World War One. When a U.S. soldier was killed in action, his life insurance payoff by Uncle Sam was sent to his next of kin. This was often a sufficient amount to pay off the mortgage on a house or farm. (At that time, a great many more people lived on farms.)
B) In the 1920s, barnstormers would travel the countryside to small cities and set up an "airplane ride" concession from some farmer's field. The pilot was expected to reimburse the farmer for any crop damage that occurred. When a pilot incurred a fatal accident he was deemed to have bought the entire farm.
C) When airports were few and far between, municipalities (or individuals with means) could sometimes persuade a farmer to sell a portion of his property to the town in question. A number of incidents occurred where much-wanted real estate would not be sold by a farmer at any cost. Since at that time, accidents were much more frequent, there came to be an association between the acquisition of some hard-won airport properties and the ultimate forfeiture of flying privileges from a fatal accident.
D) the same place that "kick the bucket" did: vaudeville.

Subject: The Pause That Refreshes

Question 16: When was the first mid-air refueling accomplished?

A) Harold Hughes, millionaire industrialist and playboy pilot, had a specially equipped Lockheed Super-Electra with a refueling hose of his own design, which in 1936 enabled him to complete an around-the-world flight in three days, 19 hours, and 17 minutes.
B) In 1931, Wiley Post and Harold Gatty flew their Lockheed Vega 5B, Winnie Mae, over the Atlantic ocean to break the around-the-world record and used a specially constructed hose for the first experimental aerial refueling off the Labrador coast.
C) Charles Lindbergh's NYP Ryan Monoplane carried 425 gallons of fuel but also had, in addition to a periscope, a telescoping pipe for aerial refueling. This was in May of 1927.
D) Stuntman and speed racer Frank Hawks and fellow daredevil pilot Earl Daugherty teamed up with wing-walker Wesley May on November 12, 1921 to complete the first aerial refueling using a five gallon can strapped to May's back as he climbed from one plane to another in midair. This inspired the US Army to develop its own air-to-air refueling program, which was later to become a critical component of military air power.

Subject: That Final Sunset

Question 17: How did the expression "going West" come to signify the death of an airman?

A) Airmail pilot Jack "Skinny" Knight flew all night through snowstorms between Chicago and Salt Lake City in a desperate attempt to secure a public relations coup in time to prevent the Air Mail Service from being cut off by the incoming Harding administration in March of 1921. Before departing, he told the airport manager to telegraph every town along the way that he was "heading west".
B) Airmail pilot Leonard Brooke Hyde-Pearson left a letter to his fellow pilots to be opened only after his death (not an unexpected event in that line of work) in which he said "I go West, but with a cheerful heart...stick to it, boys. I'm still very much with you all. See you all again."
C) Horace Greeley, ca. 1859.
D) Will Rogers, in 1935, in a remark shortly before departing on his final flight with Wiley Post.

Subject: Coffee, Tea, or Milk?

Question 18: Who was the first airline stewardess?

A) Ellen Church
B) Anne Morrow Lindbergh
C) Margaret Bourke-White
D) Jackie Cochran

Subject: The Original "Hump"

Question 19: What (and when) was the term "Hump" first significant in aviation?

A) Antarctica, 1929
B) the Sahara Desert, 1932
C) the Himalayas, 1942-1945
D) Saskatoon, Saskatchewan, 1951

Subject: The Forgotten Ancestor of the Osprey

Question 20: The concept of powered lift aircraft is not new. Aircraft such as the V-22 had a much earlier progenitor. This aircraft was:

A) an experimental ultralight aircraft, the XL-161A, which was developed in 1971, and intended for tactical interdiction. It could remain aloft for only 30 minutes, but despite the fact that there were no injuries of any kind during testing, it was scrapped.
B) a V/STOL cargo & troop transport intended for combat support. The XC-142A could carry 8000 lbs at over 400 mph, as well as in a hover. The first prototype was delivered in 1965, and the 142A made the first V/STOL landing on an aircraft carrier. However due to handling difficulties and accidents, the aircraft was cancelled.
C) the XV-3, first flown in August 1955. It had a slender, fixed, mid-fuselage wing, with large rotors on each wing tip. The rotors were vertical for take-off, landing, and hover, and rotated horizontally for forward flight, while the wing remained fixed.
D) the XV-15, which was the true "proof of concept" progenitor for the V-22, with prop-rotor engine pods mounted on wingtips, like the V-22. It first flew in October 1980.

Subject: It's The Thought That Counts

Question 21: There were, if you take it down to numerical designators, over 200 aircraft "types" either produced or used by America during World War II (although more broadly, the number of distinctly different US aircraft that saw combat is still in the dozens). How many American aircraft designs saw combat during World War I?

A) 19
B) 11
C) four
D) one

Subject: A Little More Recent Than the Caves of Lascaux

Question 22: When was the world's first numerical weather forecast made? (meaning the use of iterative techniques and atmospheric modeling)

A) in 1904, by the Norwegian Vilhelm Bjerknes
B) in 1922, by Lewis Fry Richardson
C) in 1946, by Princeton mathematician John von Neumann
D) in 1950, by ENIAC, at the Aberdeen Proving Ground

Subject: "Extree, Extree, Read All About It!"

Question 23: On September 20, 1904, Orville Wright flew in a well controlled complete circle over Huffman Prairie near Dayton, Ohio, marking the very first time that a powered aircraft had landed at its original point of departure. In which periodical did a first-hand account of this historic Wright Brothers flight first appear?

A) The West Side News, a local Dayton newspaper published by the Wright Brothers themselves, on Sunday September 25, 1904
B) the June 1905 Scientific American
C) the Smithsonian magazine, July 1905
D) the January 1905 issue of Gleanings in Bee Culture, published in Medina, Ohio by Amos Root
E) the March 1905 issue of Popular Science Monthly

Subject: The Earliest Airliner

Question 24: True or False: The first prototype airliner actually flew in the 19th century.

Subject: The "Real" Earliest Airline (with airplanes, this time)

Question 25: The first scheduled passenger airline was

A) Lufthansa
B) The Zeppelin Company
C) The St. Petersburg-Tampa Airboat Line
D) Pan Am
E) KLM

Subject: When Silver Wings And Feathers Meet

Question 26: The first reported bird strike was

A) On September 7, 1908, Orville Wright reported striking a bird (while flying) near Dayton, Ohio. It was part of a flock he was reportedly chasing. (The species was not reported.)
B) Calbraith Rogers, the first person to fly across the US, was killed on April 3rd, 1912, as a result of a bird striking his aircraft, the "Vin Fizz."
C) not until 1960, when a Lockheed Electra taking off from Boston's Logan Airport ingested a flock of starlings into three of the four engines, after which it stalled and crashed. Of the 72 passengers and crew on board, 62 died.
D) On May 23, 1931, Quentin Bodner was struck in the face by a Canada goose at Floyd Bennett Field in New York (the very same day that it was officially opened. And although it did not strike the aircraft, it was reported as a bird strike.)

Subject: Asymmetry

Question 27: Airplanes have various left turning tendencies, and designers have come up with some clever ways to counteract them. What is one that was first used during WWII, and which has hardly ever been used since?

A) a canted engine
B) a canted vertical fin
C) unequal wing lengths
D) cyclically varying propeller pitch

Subject: Early Beginnings

Question 28: Who was *really* the first person to design a balloon for lighter-than-air flight?

A) Lao-tzu, 6th century BC
B) Leonardo da Vinci, 1452-1519
C) Francesco de Lana (Lana-Terzi), 1631-1687
D) the Montgolfier brothers, 1782

Subject: Of Lines Going This Way And That

Question 29: Where and when did the present day system of latitude and longitude originate?

A) Greece, in the fourth century B.C.: The philosopher Aristotle developed it in about 335 BC in order to describe the concentric spheres in which the apparent motion of planets occurred.
B) Greece; Alexandria to be exact, and sometime in the second century, between 127 and 148 A.D. The astronomer, mathematician and geographer Claudius Ptolemy (Ptolemaeus) originated the use of lines of

latitude and longitude to better identify the location of other places on the face of the Earth. His projections were of course incomplete, being that the known world at that time was hardly truly global in extent. Rather than a complete sphere, they were more in the form of a truncated cone, extending not much further north (about 63 degrees) than the vicinity of Thule, and not very far in easterly and westerly directions.

C) from the Polish astronomer Nicholas (Nicolaus) Copernicus, ca. 1513 A.D.

D) The Portuguese map-maker Pedro Reinel was the first to draw lines of latitude, intersecting the prime meridian, back in 1506. (Lines of longitude had to await sufficiently accurate clocks!) At that time, and for the next approximately 300 years, that meridian was located in the vicinity of Madeira. The Portugese prime meridian remained until the nineteenth century, when greater numbers of nautical charts were of English origin. In 1884, a European conference "moved" the prime meridian to its present location in Greenwich, England.

Subject: Dirigible Disasters

Question 30: True or false: The Hindenburg disaster was the worst airship tragedy in history.

Subject: The Amazing Power To Disregard Those Events That Do Not Bode Well

Question 31: The historic flight of Louis Blériot across the English Channel early in the morning on July 25, 1909 did not begin favorably. Despite having done very well for himself manufacturing acetylene lamps and other accessories for automobiles, like many aviators to follow he had managed to turn a fairly large fortune into a very small one, and was nearly broke by the beginning of that year. Due to a mechanical problem during a previous flight, he had also badly burned his foot, could walk only with the assistance of crutches, and operating his rudder proved painful. In addition, the weather had been poor and there was a waiting game between Blériot and his competition. Blériot 's Model XI was puny in comparison, and was powered by an Anzani engine of only 25 horsepower. His wife Alicia did not share his love of flying and had begged him not to go. So, what else could go wrong (and did)?

A) While preparing for takeoff, a dog wandered into the arc of his propeller blade and was killed.

B) His Anzani 3 cylinder motor overheated.

C) Due to weather conditions over the channel during the flight, he could not see either coast for several minutes, and was unsure of his position.

D) He had to fight dangerous cliff-side gusts near Dover Castle and actually crash-landed when he reached the other side.

Subject: Skeletons In Aviation's Attic?

Question 32: What possibly inauspicious connection might there have been between the Wright Brothers and Lt. Thomas Etholen Selfridge, who became the first casualty of heavier-than-air flight on September 17, 1908?

A) The Army was at first disinterested in the Wright Brothers' offer, and although Selfridge, as a younger man, had wanted to work with them, they had actually turned him down. But more than that, they were somewhat jealous of Selfridge, who came from a rather well-to-do San Francisco family, and who had in 1903 graduated in the top third of his class at West Point (along with one Douglas MacArthur, who was, incidentally, the valedictorian).

B) Alexander Graham Bell, along with Canadians F. W. "Casey" Baldwin and John McCurdy (Canada's first pilot), as well as US Army officer Thomas Selfridge in 1907 founded the Aerial Experiment Association (note the similarity of its initials with a certain present-day organization) in order to implement Bell's theories of manned flight in multi-cellular tetrahedral kites. (Their funding came from Bell's wife.) But the Wright Brothers regarded Selfridge with a jaundiced eye, as he represented: the competition!

C) They were somewhat annoyed that Selfridge had made the first solo flight by a U.S. Army officer, prior to the training that was being arranged under contract with the Signal Corps, in the new 'White Wing' (the AEA's second airplane) on May 19, 1908 at Hammondsport, NY.

D) There may be some truth to all three, but most of all, there was one more member of the AEA that I didn't mention in choice B: Glenn Curtiss. Orville Wright, who was the pilot of that fateful flight, was suspicious of Lt. Selfridge because of his close association with Curtiss, who had, the Wright Brothers believed, recently infringed upon their patent.

Subject: First Trap

Question 33: The first aircraft to land aboard a ship was:

A) on January 11, 1928. The USS Saratoga, the first carrier in the United States Navy, sailed from Philadelphia for her shakedown cruise on January 6. Five days later, her air officer Marc A. Mitscher, future World War II hero (and Admiral), 1910 graduate of the US Naval Academy, and 1916 graduate of the Navy's first formal flight school at Pensacola (Naval Aviator #33) landed the first aircraft on board a carrier.

B) earlier: on October 26, 1922. Lieutenant Commander Godfrey de Courcelles Chevalier, flying an Aeromarine, made the first landing aboard USS Langley underway off Cape Henry, Virginia. (He died less than one month later in the Naval Hospital at Portsmouth, VA from injuries suffered in a plane crash near Norfolk.) The first carrier take-off had just been made nine days earlier (with a Vought VE-7SF, flown off the Langley, but by another pilot). The Langley had been commissioned just over a half-year earlier at Norfolk, having been converted from the collier USS Jupiter as the Navy's first aircraft carrier. The engineering plans for this conversion included catapults fitted on both forward and after ends of the deck.

C) on January 3, 1918. During World War I, Great Britain's Royal Navy refurbished a partially completed cruiser, equipped with horizontal smokestacks like exhaust pipes, a hydraulically-mounted wheelhouse (designed to retract under the deck when launching airplanes), and initially tested on that date. Renamed the HMS Argus, it wasn't ready before the end of WWI, but it was the world's first" through-deck" aircraft carrier.

D) on Jan. 18, 1911, at 11:01 AM, to be exact.., Eugene Ely, a civilian pilot flying a Curtiss pusher, landed on a custom-built platform on the cruiser USS Pennsylvania, at anchor 13 miles offshore in San Francisco Bay. A few months earlier, in November 1910, he had become the first person to take off from a ship (the USS Birmingham, at anchor in Hampton Roads, VA). As his Curtis Biplane swooped down toward the stern, both pilot and crew waited anxiously, hoping that the hooks on his plane's underside would snag the crude rope and sandbag rigging that was strung across the deck. It worked, and carrier aviation was born...well, sort of.

Subject: Early Birds

Question 34: The first wartime aerial photo reconnaissance was used

A) During the War of 1812
B) During the Civil War
C) During WWI
D) During WWII

Subject: The Early Bird Gets the...Patent?

Question 35: How long after their first flight did the Wright brothers submit their first patent application?

A) six years
B) three years
C) three months
D) three days

Subject: The First Composite Airplane

Question 36: In what year was the first commercially built composite aircraft flown?

A) 1983, Beech Starship
B) 1980, Glasair
C) 1967, Windecker Eagle
D) 1938, Short Mayo

Subject: The REAL "World's First Aircraft Carrier"

Question 37: An earlier trivia question featured the so-called birth of carrier aviation with the world's first landing aboard a ship. So, what was *really* the world's first aircraft carrier?

A) The USS Saratoga, although it was completed in 1927, became the first operational aircraft carrier of the United States Navy to be originally built for that purpose, in January 1928.
B) New times required new aircraft types, and the world's first aircraft carrier belonged to the United States Navy. However, it was even earlier. In this case, it was the USS Langley, which had been converted from the collier (term for a ship used to carry coal) USS Jupiter as the Navy's first aircraft carrier, back in 1922.
C) If a ship carries airplanes, and airplanes can take off from them, then it's an aircraft carrier, by definition if not by formally declared class. Dismissing the philosophical argument of whether the implied criterion of also being able to return to the roost must be considered, the point at which a ship first became an aircraft carrier was in November of 1910, when an airplane took off from the USS Birmingham.
D) Actually, extending definitions a bit further, the world's first aircraft carrier dates back almost half a century earlier: to November of 1861.

Subject: So Much For the "Big Sky" Theory...Even Then

Question 38: The first mid-air collision (between two aircraft) occurred

A) Would you believe 1910? And would you believe that the two pilots were brothers?
B) in 1914, on August 26, during the first air battle on the Eastern Front during World War One. It was also, somewhat less than nobly, the first *deliberate* mid-air collision. Staff Captain Petr Nikolaevich Nesterov scored an aerial victory, in this case also the first ever of its kind, by ramming. It is likely that he performed this maneuver, called "Taran", our of desperation because he ran out of ammunition. In this case, he ran his propeller into the tail of an Austrian Albatross bomber. It was a Pyrrhic victory, however, because he did not survive it. (He was also the first pilot to perform an inside loop.)
C) in 1919, between the Navy dirigible DN-1 and the Goodyear dirigible FB-1---yes, there was a "Goodyear blimp" of sorts, even back then, although it was acquired under Navy contract---over Akron Ohio. However they were both traveling very slowly, it was a glancing impact,

and they each basically just bounced off the other, and went on their merry way.

D) on April 21, 1929, when a Ford Tri-Motor operated by Maddux Air Lines was struck over San Diego by a single-engine Boeing B-PW-9D pursuit biplane being flown above the airliner by an Army pilot, who decided to turn steeply down and pass in front of the other airplane. Instead, he misjudged the maneuver and slammed into the cockpit of the Tri-Motor at an altitude of 2000 feet. There were no survivors.

Subject: First Blood

Question 39: The first use of an airplane for intentional civil violence (pardon the oxymoron) was

A) in 1964. on May 7, an intruder shot the pilot of the Pacific Airlines Fairchild F-27, which then crashed near Doublin, CA, killing all aboard.
B) in 1955. On November 1st, a United Airlines DC-6B exploded in flight and crashed near Longmont Colorado. Investigators later discovered that someone had placed a bomb aboard the airplane in order to collect on an insurance policy taken out on a passenger (which in this case was that person's mother).
C) on November 12, 1926. Three bombs were dropped on a farmhouse in Williamson County, Illinois. In this case, it was a feud between two rival bands of bootleggers, the Shelton gang, and the Birger gang. Regrettably, none detonated.
D) on November 1, 1911. An Etrich Taube monoplane was used by Italy to drop bombs in Turkey.

Subject: The Original Genius?

Question 40: Who was the first person in history to write about flying in any mechanical or truly scientific sense?

A) Leonardo da Vinci (1452-1519)
B) Roger Bacon (1214 - 1294)
C) Abbas ibn Firnas of Islamic Spain, in the 800's A.D
D) the Greek, Archytas of Tarentum, ca. 428-350 B.C.
E) no one really knows

Subject: Too Good To Be True

Question 41: Before 1900, there was a "law" of aerodynamics, derived experimentally by the renowned Samuel Pierpont Langley, which said that the total power required for flight decreases, as speed increases. While his data were not in error, his methods were still flawed. Why?

A) He failed to account for parasite drag.
B) He used mathematical modeling techniques to apply data taken from smaller-scale models, but he did so incorrectly.
C) He used a limited range of velocities, none of which exceeded the minimum on the curve of power vs. airspeed.
D) He used airspeeds which were too high.

Subject: Hey, Didn't the Wright Brothers *Ever* Smile?

Question 42: The Wright Flyer was not displayed at the Smithsonian until:

A) 1912
B) 1924
C) 1936
D) 1948

Subject: Pea Soup

Question 43: The first fully automated landing of a passenger-carrying commercial aircraft was in

A) 1929
B) 1965
C) 1977
D) none of the above

Subject: The REAL "first hot air balloon"

Question 44: In what year did the first flight of a hot air balloon occur?

A) 1519
B) 1602
C) 1709
D) 1783

Subject: Photo Op

Question 45: The first photographed supersonic flight didn't occur until

A) 1949
B) 1947
C) 1942
D) 1887

Subject: Q Branch, and the Q Quandary

Question 46: What was (is) the significance of QFE and QNH in aviation?

A) none. These were early telegraphic codes used in the maritime services
B) absolutely none. These were Ian Fleming's two fictional branches for spy gadgetry in the original 13 James Bond novels.
C) altimetry
D) transponder operation

Subject: The Early Bird Gets the...Patent? ...Part II

Question 47: Approximately how many patents for flying machines were filed with the US Patent Office *before* that of the Wright Brothers?

A) approximately 400
B) about 175
C) roughly 20
D) exactly none

Subject: Got Wheels?

Question 48: The first airplane to take off on wheels was

A) the White Wing, Glenn Curtiss, 1908
B) The Bleriot XI, Louis Bleriot, 1909
C) Wright Flyer, modified by Lt. Benjamin D. Foulois, 1909
D) Wright Flyer, Model B, 1910

Subject: A Hot Idea

Question 49: Who invented the first thermometer?

A) Guglielmo Marconi
B) Daniel Gabriel Fahrenheit
C) Galileo Galilei
D) Aristotle

Subject: The First Jet Carrier Ops

Question 50: The first airplane to land aboard an aircraft carrier under jet power was

A) the McDonnell-Douglas XFD-1 Phantom, in July of 1946
B) the de Havilland Vampire I, in December of 1945
C) the Ryan FR-1 Fireball, in November of 1945
D) the He-178 in August of 1939

Subject: The First Seaplane

Question 51: The world's first seaplane was built by:

A) Henri Fabre, in 1910
B) Glenn Curtiss, in 1911
C) TOM Sopwith, 1912
D) the Wright Brothers, their Model CH, in 1913

Subject: Skin and Bones

Question 52: Who invented the monocoque fuselage, and when?

A) Dr. Adolph Rohrbach, 1924
B) Glenn Hammond Curtiss, 1918
C) Igor Ivan Sikorsky, 1913
D) Louis Béchereau, 1911

Subject: This Is No Tale

Question 53: Who invented the Flying Wing type aircraft?

A) Howard Robard Hughes, Jr.
B) no one individual can claim full credit
C) Hugo Junkers
D) John Knudsen Northrop

Subject: All In Knots

Question 54: When did our primary airspeed indications (that is, those used by general aviation) go from reading in miles per hour to being given instead in terms of knots?

A) on June 26, 1946
B) They never really did; for civil aircraft, there was never any rule absolutely requiring it, so whatever you see in whatever aircraft you happen to fly depends only upon the particular manufacturer, and the caprices of fate.
C) in 1977
D) For most general aviation aircraft, it was on January 3, 1991, with Amendment 23-42 to CFR Title 14, FAR Part 23.1323.

Subject: Where the Force Started

Question 55: The US Air Force originated in

A) 1907

B) 1918
C) 1926
D) 1947

Subject: Aces High

Question 56: Arrange in their proper respective order the approximate numbers of American aces in the First World War, World War II, Korea, and Vietnam.

A) 100, 1300 ,50, 3
B) 30, 150, 150, 30
C) 60, 600, 150, 15
D) 20, 100, 500, 20

Subject: Broken Arrows

Question 57: The United States has (so far) been the only nation to drop an atomic weapon on another country. As everyone knows, we dropped two atomic bombs on Japan during World War II (and it was this which, in fact, hastened the end of that war). Okay then...

True or false: The United States has never dropped the more powerful *hydrogen* bomb on any country--either during times of peace, or while we were at war.

Subject: Unfriendly Skies

Question 58: The first aerial combat between two aircraft took place

A) between an English and a German dirigible over France, in 1912
B) between two Americans over Mexico in 1913
C) near the beginning of WWI, in 1914, between two unidentified observation planes
D) on April 1, 1915, when the Jean Navarre brought down his first enemy aircraft with a rifle (This was also the same date that Roland Garros scored his first victory.)

Subject: Falling Kangaroos

Question 59: What's the safest airline in the world?

A) AirTran
B) Qantas
C) El Al
D) Hawaiian Airlines

Subject: Founding Fathers

Question 60: Who was the first person to propose the use of aircraft in the national interest?

A) the Wright Brothers, in 1905
B) Glenn Curtiss, in 1910
C) Thaddeus Lowe, in 1861
D) Thomas Jefferson, in 1784

Subject: Insignia

Question 61: If you saw this on a military aircraft, what would it tell you?

A) that you're looking at either a Navy or Marine Corps aircraft
B) that it once belonged to the Japanese army
C) that it's a very *old* aircraft
D) that you're looking really close, because this is the insignia used on aircraft used for clandestine operations, and it's only one inch tall

Subject: Friends In High Places

Question 62: Of the following early airmail pilots, who doesn't belong?

A) Charles Lindbergh
B) Elrey Jeppesen
C) Leonard Brooke Hyde-Pearson
D) George H. Boyle

Subject: Blind Ambition

Question 63: The first person to make an intentional "blind" landing was

A) Omar Bradley
B) Wesley L. Smith
C) Jimmy Doolittle
D) Arthur Godfrey

Subject: ROGER: We have clearance, Clarence.
CLARENCE: Roger, Roger. What's our vector, Victor?

Question 64: Just where and when did pilots start saying "Roger" to acknowledge receipt of a radio communication, anyway?

A) College Park, Maryland, 1911, at the Army Aviation School
B) Washington D.C., in 1946, by the Bureau of Aeronautics
C) in the United States, since before WWII
D) only since the Korean War

Subject: Honorable Mention

Question 65: Chicago's O'Hare Airport is named after

A) the first licensed pilot in the state of Illinois, Desmond O'Hare
B) the son of a mobster who worked for Al Capone
C) Teddy Roosevelt's eldest daughter
D) the brother of Kate O'Leary (whose cow supposedly started the great Chicago fire of 1871)

Subject: Humble Beginnings

Question 66: Embry-Riddle Aeronautical University in part owes its start to

A) a barnstormer and an entrepreneur
B) the Weaver Aircraft Company (a.k.a. Waco)
C) a flying squirrel
D) a moose

Subject: Not *the* First

Question 67: Charles Augustus Lindbergh was not the first person to fly across the Atlantic. He was

A) the 13th
B) about the 27th
C) somewhere between the 85th and the 92nd
D) the 99th

Subject: Big Budgets

Question 68: What was the approximate total out-of-pocket cost to Wilbur and Orville Wright in developing their Flyer, including travel costs and materials?

A) $17,000
B) $42.50

C) $850
D) $66,000

Subject: Trailblazer

Question 69: Which airline was the first in the Western world to hire a woman pilot?

A) American Airlines
B) Eastern Airlines
C) Central Airlines
D) Western Airlines

Subject: Collector's Item

Question 70: The first US postage stamp featuring an aircraft was issued in:

A) 1917
B) 1912
C) 1908
D) 1889

Subject: How Old Is the Attic?

Question 71: When did the Smithsonian Institution first begin collecting flight-related articles?

A) No aviation artifacts were owned by the United States National Museum (its name, before 1911) until the year 1910.
B) Not until 1948, with the acquisition (somewhat belatedly) of the Wright Flyer.
C) It actually began with the 1931 deposit of the Langley Aeronautical Collection.
D) A small selection of kites were donated by the Chinese Imperial Commission in 1876.

History Lessons:
Answers

Answer to Question #1: The answer is choice D. Units that stopped using tracers saw their success rate nearly double and their loss rate go down.

Answer to Question #2: The answer, somewhat disappointingly, is actually C. Choice D is urban legend, although some versions have that name as "Gorsky". The words in choice C were then followed by "ACA out of detent"...and a bunch of other stuff before, 19 seconds later, the words "Houston, Tranquility Base here...") Also they were spoken by the Lunar Module pilot (who I believe was Neil Armstrong, and not Buzz). This information comes from the Apollo 11 Technical Air-To-Ground Voice Transmissions, compiled by the Data Logistics Office of the Test Division of the Apollo Program Office, in July of 1969.

Answer to Question #3: The answer (of course) is choice A. The dirigible crossings were made in 1919, but it wasn't the first non-stop crossing by air. Alcock and Brown were first. (And incidentally, the first crossing of the even *wider* Pacific Ocean was just four years after Lindbergh, between October 4 and 5 of 1931. Hugh Herndon and Clyde Pangborn made the first nonstop transpacific crossing, going from Sabishiro Beach in Japan to Wenatchee in Washington state, in about 41 and a quarter hours.)

Answer to Question #4: The answer is choice B. Choice C is close, as Lindemann was more of a theoretician (and he was a scientific advisor to Winston Churchill in later years, as well as a viscount). However, Parke was first. He enjoyed this distinction for all of two months, however, before he died trying to return to the runway after a low-altitude engine failure. (See? Slay one dragon, get clobbered by another...) The person credited with actually figuring out how to get out of a spin however (with Parke, it was most likely a matter of luck) was Lindemann (choice C).

Answer to Question #5: The answer is choice D.

Answer to Question #6: This is a toughie. Depending upon which YOU think is worse—being soullessly duplicitous or calamitously misguided, either C or D would count as being correct. My money's on choice C. If I knew where his grave was, I'd go spit on it.

Answer to Question #7: It depends on what you mean by "to fly". If you mean a powered, sustained, and fully controlled flight from level ground, followed by a climb and descent to ground no lower than what you'd taken off from, well then the commonly accepted answer is probably Orville Wright on December 17, 1903. But Clement Ader, a Frenchman, designed and built a bat-like machine powered by a steam engine driving a tractor propeller. It lacked any kind of control surfaces and was unable to maintain sustained flight (though it did travel some 160 feet). Felix du Temple's aircraft became airborne only after plunging down a steep ramp. The last one is actually the most amazing, because of its surprisingly early date. Sir George Cayley's machine did have control surfaces, but they had to be "set" before flight (not much good there). Also, of the four choices, it was the only one that was non-powered. After several short glides using only ballast, Sir George delegated this epic achievement* to the son of a worker from his estate near Scarborough, England; his identity remains unknown. In 1853 also, his coachman made a longer flight across an entire valley on the estate. (The more comical record of history was that upon landing, the man quit his post on the spot, saying that he was not being paid to fly.) So, I say it's that unsung lad, and that it's choice D.

Answer to Question #8: Actually, airborne armament was used well before WWI, and the answer is choice A. As early as June 1912, the US Army began trials using machine guns mounted in a Wright Model B biplane, and on June 30, 1910, Glenn Curtiss had dropped dummy bombs on an outline of a battleship buoyed on Lake Keuka, in New York. The aviation industry in those years really wasn't set up for mass production, and so for military purposes and standardization of equipment, as well as tactical applications across squadrons, things began slowly. The primary task for aircraft had started with reconnaissance from free balloons. It was soon realized of course that it would be possible to keep the enemy from acquiring information by shooting him down (or otherwise forcing him back to earth). Also, the relationship between the machine gun and the propeller did at first cause some technical difficulties...

Answer to Question #9: Commercial transport of passengers began in earnest immediately after World War I. They primarily served short-term political purposes, not economic ones, by allowing diplomats and politicians to more expeditiously sort out the political and financial legacies of the "Great War". The routes were technically open to the public, but the passengers would have to have primarily been wealthy joy-riders, interested mostly in the novelty of transportation by air rather than its (eventual) utility. All that said though, on the first day of the year 1914, the Benoist Company scheduled passenger service from Tampa Florida to St. Petersburg, Florida. The fare was five dollars. The Benoist Type XIV "Air Boat" made its first flight over Tampa Bay (25 miles, at 60 mph) on January 12, 1914. The correct answer is choice A.

Answer to Question #10: The answer is choice C.

Answer to Question #11: The answer is actually choice D. An itinerant entrepreneur named Peter Carnes gave a public demonstration flight in Baltimore in 1784. A 13 year old boy named Edward Warren volunteered his services as the first passenger. He ascended to approximately 200 feet (on a tethered line).

Answer to Question #12: A. The Supre Sabre, the first of the F-100 series, known by those who flew it as simply "the Hun", was in service mostly in the 1950's (although it did see action in Vietnam, and the Thunderbirds used it until 1964). It could not exceed Mach 1 at sea level, but its top speed at 35,000 feet was over 850 mph (and so the right answer is choice A). It was designed to replace the F-86 Sabre. The F-101 (choice B) was the first airplane to exceed 1000 mph in level flight. Although there was an F-103 in design, none were built. The F-104 (choice D) was the real star of the century series, performance-wise, reaching over 1400 mph and altitudes over 100,000 feet as early as the late 1950s.

Answer to Question #13: C (shhh!) The Quiet Birdmen was founded by seven World War I pilots who, beginning in 1919, met regularly met at Marta's Restaurant in Manhattan to share war stories. In the spring of 1920 the group, which had grown considerably, rented the roof garden restaurant on top of a Manhattan hotel populated principally by elderly residents. They kept the roof for the entire weekend during which they partied quite heartily. They created quite a ruckus on the roof and while going up and down the stairwells and the elevator. Due to the noise, the elderly residents dubbed the group "The Quiet Birdmen". The name stuck and "Ye Anciente and Secret order of Quiet Birdmen" was officially founded in January 1921.

Answer to Question #14: FALSE! We know *approximately* where, of course. Kill Devil Hill, one of several sand hills near Kitty Hawk, North Carolina, is like any other sand dune: a few decades go by, and wind and storms take away what was, so that the shifting sands look nothing like they did over 100 years ago. In 1928 William Tate, the gentleman with whom Wilbur Wright stayed when he first visited Kitty Hawk, and three of the several men who had witnessed the historic flight (and were still living) got together to try and figure out exactly where that was, so that a monument could be built. Based on the remains of a storage building where the Wright Flyer had been kept, and their collective memories that the 60-foot track used to launch it had been about 100 feet North of that building, they marked that spot---a quarter century later---with a copper pipe. (This was soon replaced with a six-foot stone and bronze tablet.)

Answer to Question #15: It's choices A and B.

Answer to Question #16: It was choice D, the earliest (and compared to the others, most unknown) of them. Though their other accomplishments are well known, any parts involving midair refueling were total fabrication, and none of the others made any pioneering efforts in this area.

Answer to Question #17: It is choice B. The pilots who flew the early airmail routes had a seriousness and a sense of commitment in an era of trial-and-error technology which set them completely apart from the itinerant barnstormers, and their character and philosophy reflected the period in which they lived--and died. The high regard held by pilots flying for the early Air Mail Service was poignantly revealed in the early 1920s by the contents of a letter addressed to the Service by one of its pilots, Capt. Leonard Brooke Hyde-Pearson. His mother released this letter to the Post Office Department shortly after being notified that her son had crashed and died in a Pennsylvania forest, during a blinding snowstorm. It read as follows: "To My Beloved Brother Pilots and Pals: I go west, but with cheerful heart. I hope whatever small sacrifice I have made may be of some use to the cause. When we fly we are fools, they say. When we are dead, weren't half-bad fellows. But everyone in this wonderful aviation service is doing the world far more good than the public can appreciate. We risk our necks; we give our lives; we perfect a service for the benefit of the world at large. They, mind you, are the ones who call us fools. But stick to it, boys. I'm still very much with you all. See you all again." The circumstances before and during Jack Knight's flight (choice A) were real, and in fact it had a happy outcome. But it was actually between Omaha and Chicago, and his telegraph request was actually for lights and bonfires to help guide him through the foul weather.

Answer to Question #18: It is choice A. She persuaded the president of Boeing that female crewmembers would convince reluctant fliers that air travel was safe. In 1930 she and seven other registered nurses became the first stewardesses (now of course, known as flight attendants). Anne Morrow Lindbergh wrote many classic memoirs about flights with her equally famous husband, as well as many other things, but never served anyone other than her husband. Margaret Bourke-White was a famous photographer, and Jackie Cochran formed the Women's Auxiliary Flying Squadron in addition to having made many other aviation firsts. (Serving coffee wasn't one of them however.)

Answer to Question #19: It is choice A. It was the pass in the Queen Maud Range between Little America and the South Pole that Admiral Richard Byrd, Bernt Balchen, Harold June, and Ashley McKinley had to hurdle in the Ford Trimotor used during Byrd's Antarctic Expedition, in 1929 (the lowest elevation of which was approximately 10,500 feet). This was the first "hump" one will find referenced in aviation literature. Choices B and D are decoys. Choice C of course is what most people would think of first. World War II transport pilots rightly viewed that "hump" as the most treacherous air route in the world. After the occupation of China by the Japanese, the only link between India and China was a hazardous air route across the rugged Himalayan Mountains--the famed "Hump." The obstacles posed by terrain, extremes in climate, and attacking enemy planes all combined were difficulties never before experienced in mass operations of aircraft. Across this route, the US Army Air Force undertook and maintained the aerial re-supply of China in the greatest sustained aerial transport achievement of WWII, carrying cargo ranging from bombs, gasoline, and medicine to spare parts, trucks, and K-rations. Flights began in April 1942 when the US Army flew gasoline and oil to China for planned use by Doolittle's Raiders following their attack on Tokyo. The India-China Wing of the Army Air Force's Air Transport Command slowly increased its lift over the "Hump" to more than 12,000 tons a month in early 1944 and 71,000 tons in July 1945. It cost the lives of some 800 flyers. The mission required thousands of flights over the highest mountain range in the world to provide more than one million tons of cargo and relief supplies to friendly forces operating in China and Southeast Asia.

That was a most noble and mighty effort, indeed. However, the crew in Admiral Byrd's expedition also had quite unique, unusually harsh, and other-worldly hardships to deal with. On Thanksgiving Day in 1929, pilot Bernt Balchen and his crew aboard the "Floyd Bennett" (after Byrd's late friend, with whom he had flown over the North Pole, three years earlier) flew south from the ice pack at 90 mph and climbed to 8,000 feet. As they approached the Queen Maud range, they were forced to jettison everything, (including some food and emergency supplies) so that they could clear the glacial summits and reach the polar plateau. At 1:14 a.m. the following morning, Byrd and his men reached the Pole. (Byrd was the first to fly over both North and South poles, though there is some latitude—hah, hah—regarding the former.)

Answer to Question #20: Choice C is the right one.

Answer to Question #21: It was choice D. Our "Jenny" or "Standard" trainers were never used in combat, but the Curtiss Flying Boat was used for submarine patrol and bombing. (It must be mentioned however that the United States didn't formally decide to help make the world safe for democracy until April 6, 1917, so the other Allies had a bit of a head start...)

Answer to Question #22: The right answer is choice D. Numerical weather prediction by integrating the governing hydrodynamic and thermodynamic equations forward in time, starting from the observed, initial state of the atmosphere, was proposed in 1904. Richardson developed the first numerical weather prediction (NWP) system. His technique of dividing the atmosphere into grid cells, then using finite difference solutions of differential equations, formed the basis for future efforts. Soon after ENIAC (which was the world's first electronic computer, and stood for Electronic Numerical Integrator And Computer) became operational, in 1946, von Neumann advocated the application of computers to weather prediction. The first computerized weather forecast was run on the ENIAC in 1950. The model used a two-dimensional grid with 270 points about 700 km apart, and a time step of three hours. A few years later, the

Weather Bureau and several research and forecasting agencies of the Air Force and Navy established a Joint Numerical Weather Prediction Unit in Suitland, Maryland. It began routine real-time weather forecasting in May, 1955. However, it was well over a decade before numerical methods became more accurate than the "subjective method" of barotropic and baroclinic models (and educated guesswork) used by human forecasters.

Answer to Question #23: It's choice D. Although the Wright Brothers had three careers (publishing, bicycles, and then aviation) they had left the first one behind before their well known exploits in North Carolina as well as back home in Ohio. (Actually, while they we involved with the Wright Cycle Company, they did advertise that fact in a weekly magazine that they printed.) The fact is though that the Wrights' own press release about their circular success, which was sent out the following month, was simply ignored. This might have been because many people just didn't believe it, even though there were witnesses to their first flights. A report by Octave Chanute did appear in the March 1904 issue of Popular Science Monthly (not 1905). However this was about their earlier accomplishments of course, not this one. Perhaps because of sheer disinterest, the relative isolation of the times (or in the case of the latter, rivalry with Samuel Langley) neither Scientific American nor Smithsonian got the scoop. In fact, it was someone else: Amos I. Root, the owner of a beekeeping supply house in Medina, Ohio (and also the inventor of the modern beehive), had heard that men were actually flying at Huffman Prairie, and being somewhat of an early adapter, he motored 175 miles to Dayton--I don't know what type of automobile it was--to see for himself. He was on hand to see Orville's triumphant flight, and he later published the first eyewitness account of a sustained, controlled, powered flight that involved a return to the original takeoff point in the January 1905 edition of all things, Gleanings in Bee Culture, a journal he published for his customers. (Root later offered to allow Scientific American to reprint his articles, but they refused.) The editors of Scientific American no doubt felt stung (sorry), having missed out on such a story. He was so impressed by what he saw that he wrote and published a series of articles in this publication (starting in January, 1905) for the next two years, reporting developments from Huffman Prairie. His writings were eloquent, accurate and very supportive. So the best coverage of the development of the airplane was written for an audience of beekeepers! Incidentally the magazine (now known as Bee Culture) is still published today in Medina OH. Here is Amos Root's first report:

> "When it turned that circle, and came near the starting-point, I was right in front of it; and I said then, and I believe still, it was...the grandest sight of my life. Imagine a locomotive that had left its track, and is climbing up in the air right toward you -- a locomotive without any wheels...but with white wings instead...spread 20 feet each way, coming right toward you with a tremendous flap of its propellers, and you have something like what I saw...I tell you, friends, the sensation that one feels is something hard to describe."

...and in case you're wondering, one of the first known public announcements about the first and most historic of their flights (the very first one on December 17, 1903) appeared the very next day in the Dayton Daily News of December 18, 1903, the headline for which was: "Dayton Boys Emulate Great Santos-Dumont".

Answer to Question #24: TRUE! The world's first passenger airline was DELAG (a contraction which stood for Deutsche Luftschiffahrts AG, or German Airship Transport Company), and was formed in 1909. As you might guess, these were not airplanes, but airships. There were aerodromes in Berlin, Dresden, Frankfurt, and Hamburg. By 1913 they had made more than 1,600 flights in perfect safety, carrying over 35,000 passengers. They also became the first airline to hire a flight attendant. His name was Heinrich Kubis, hired in March 1912, and he continued as a steward in zeppelins until 1939. He was chief steward on the Hindenberg, in fact (and yes, he did survive). Some of you may remember an earlier trivia item about another "first" cabin attendant, the "first stewardess" Ellen Church, and which wasn't until 1930.

But the first trial of a prototype airship which would later enter commercial service was in fact technically in the nineteenth century, on July 2, 1900. These were known as zeppelins of course, in honor of Count Ferdinand von Zeppelin, who supplied them. The early prototype was about 420 feet long and 38 feet wide, used hydrogen, was powered by two 15 horsepower Daimler engines, each of which drove two propellers, and it carried just five people on that flight. That first flight was only at a modest 1300 feet and covered less than four miles.

Answer to Question #25: The correct choice is C. Deutsche Luftreederie (later, Lufthansa) was the first operational airline (using airplanes) providing sustained passenger service. It began service flying from Berlin to Leipzig and Weimar on February 5, 1919 (followed only three days later by the French Farman Company, crossing from Paris to London). However, all that said, on January 1, 1914, the St. Petersburg-Tampa Airboat Line became the first airline to provide a scheduled air passenger service. Although it lasted only three months, the company flew over a thousand passengers across an eighteen-mile stretch of water at a cost of five dollars per passenger. KLM also comes close though, and is one of the (if not the) oldest airlines still flying under its original name. Founded on October 7, 1919, KLM (an abbreviation of Koninklijke Luchtvaartmaatschappij NV, which in Dutch means Royal Air Transportation Company), it flew its first scheduled service, between Amsterdam and London, on May 17, 1920.

Answer to Question #26: choice A

Answer to Question #27: It's choice C. The Macchi Mc.202 Folgore was designed by Mario Castoldi over 60 years ago with a feature that one would expect of Burt Rutan (and in fact which he has used on Proteus): the left wing is longer than the right (in the case of the Folgore, by about eight and a quarter inches). Choices A and B are quite common; choice D is common, too, but on helicopters.

Answer to Question #28: Choice C is correct. In 1670, an Italian Jesuit (and professor of physics and mathematics) by the name of Francesco de Lana studied the principles of the balloon and designed a ship consisting of four spheres as depicted in this woodcut, from each of which the air was to be removed, and which would be attached to a basket. He correctly deduced that if the air displaced by the balloon is heavier than the balloon, then the balloon would fly. Of course, no vessel could be constructed, then or now, that would be both light and strong enough (even one of spherical shape) to lift even itself alone by means of a contained vacuum.

Answer to Question #29: The right choice is B. Ptolemy published his *Geography*, consisting of several volumes, which became the standard textbook on the subject until the 15th century. (Copernicus first wrote on his heliocentric theory in 1513.) Such lines were more conceptual than practical, of course, so a case *could* be partially made for choice D!

Answer to Question #30: FALSE. On October 5, 1930, while the British airship R101 was en route to India, it encountered a heavy rainstorm and crashed into a hill at Allone near Beauvais, France. At the time it was the world's largest airship, 777 feet in length. It was destroyed by fire, and at least 46 passengers lost their lives. Its sister ship the R100 was scrapped after the destruction of the R101, and British government abandoned construction of all dirigibles.

On April 3, 1933, the USS Akron cast off from the Naval Air Station at Lakehurst NJ for a routine training flight. The ship was overtaken by a severe cold front. The Akron turned eastward to ride out the storm at sea and just after midnight, the Akron was caught in turbulent downdrafts. With the altimeter reading eight hundred feet, the ship lurched suddenly. The crew assumed it was a gust, but it was actually the lower fin hitting the water; no one realized the severity of the low pressure system through which the Akron was flying. The Akron had literally flown into the sea. The Akron carried no life jackets and only one raft. Most of the men never got out of the foundering ship, and of those who did, only three survived the chilly North Atlantic waters. Seventy-three men perished. (The tragedy was compounded the next day when another blimp crashed while looking for survivors, losing two of its crewmembers.)

Over four years later, on May 3rd, 1937, the Hindenburg departed Frankfurt Germany for Lakehurst, on the first of many planned flights to the US. Aboard ship were 36 pampered passengers and a crew of 61. Delayed by persistent head winds, it didn't reach Lakehurst until late afternoon on the sixth. At 7:25 PM, hundreds of people were there to watch it land, and a radio announcer was describing how beautiful it looked against the night sky. Suddenly a tongue of flame appeared and the ship was consumed in a huge ball of fire, crashing into the ground only half a minute later. Thirty-six people died: 13 passengers, 22 crewmen, and one civilian member of the ground crew. Whether it was sabotage to discredit the Nazi regime, or the highly flammable hydrogen was ignited by a spark, or possibly an electrostatic discharge from the substance used to "dope" the cotton skin, it was a terrible tragedy. Nonetheless, that of the R101 was greater, and the USS Akron was, by pure numbers, twice as unfortunate.

Answer to Question #31: Any and all are correct. (Rain showers cooled the motor, probably enabling him to complete the crossing.) His fortunes did later change for the better, though. Aside from having won the sought-after London Daily Mail prize of 1000 pounds, during the First World War his company produced the famous S.P.A.D. fighter, and "the father of the modern monoplane" also played a role in developing commercial aircraft after the war.

Answer to Question #32: choice D, of course. Selfridge did fly with the AEA several more times, but this was before he was actually assigned to the Signal Corps' Aeronautical Division, at Fort Myer, Virginia. (It was actually Selfridge who convinced Bell to move to Hammondsport, New York, where Glenn Curtiss was building a 40-horsepower, 8-cyclinder gasoline engine.) History often has twists of fate, and this event certainly had its share. In September of 1908, Orville went to Fort Myer to demonstrate the Flyer to the Army. Selfridge had made arrangements to ride along with him on September 19, but then he received orders to go to the Missouri State Fair and Military Tournament for a Dirigible exhibition (which meant that he would lose his chance for the flight). However, at the last minute, another officer traded him for that flight. That morning, Orville flew alone around Fort Myer, but he wasn't satisfied with one of the propellers. He replaced it with a newer, longer version in time for the flight with Lt. Selfridge. (The new type of propeller had never been flight-tested.) It is believed that prior to that fateful flight, Orville had (most ominously) written to Wilbur saying "I will be glad to have Selfridge out of the way." Little did he know. The Flyer circled Fort Myer several times at about 150 feet. During their fifth time, there was a loud bang and one end of the propeller blade broke off. To bring the Flyer under control, Orville shut down the engine and was able to glide to about 75 feet, but then it nose-dived into the ground. Selfridge died three hours later, and Orville was hospitalized for three months.

reproduced with permission from First Flight Shrine, Wright Brothers National Memorial and National Park Service - Outer Banks Group

Incidentally, although he was the first person to die in an aviation accident, Selfridge was not a pilot at the time. The first pilot to die (while at the controls of an airplane) was Eugène Lefebvre, on September 7, 1909, while testing a new French-built Wright airplane. The first American pilot fatality was a man on the Wright brothers' exhibition team, Ralph Johnstone, from Kansas City. A born daredevil, he had earned a living as a trick bicycle rider since the age of 15, doing seemingly impossible things such as mid-air flips from a springboard (while riding). On November 17, 1910 however, while performing a spiral dive at an air meet in Denver, he failed to pull out (possibly due to mechanical causes).

Answer to Question #33: choice D. If you took the question to mean landing aboard an aircraft carrier proper, then you might insist that it should be choice B. (The definition of an aircraft carrier itself obviously changed as both designs and operational techniques were refined over the years, but I very deviously did not commit to that term, specifically.) The photograph on the left was taken immediately after Eugene Ely landed on a custom-built platform on the cruiser USS Pennsylvania, at anchor 13 miles offshore in San Francisco Bay on Jan. 18,

1911. The one on the right shows Lieutenant Commander Godfrey de Courcelles Chevalier making the first landing aboard the USS Langley underway off Cape Henry, Virginia on October 26, 1922.

photographs courtesy U.S. Naval Historical Center

Answer to Question #34: choice C. Most of the recognition and progress in aerial photo reconnaissance generally shows up in the history books beginning with World War One. Although some clever innovator probably made the connection, and it is likely that photography probably took to the air fairly soon after it was invented in the mid-nineteenth century, no vertical or oblique aerial photography captured by balloons from the Civil War period have been found. During the Civil War, there certainly was a US Army Balloon Corps, but most activity (if not all) was limited to verbal descriptions and sketches. Despite its advantage to the Union Army, the unit was deactivated in 1863. (They had a not-surprising tendency to draw enemy fire.) Credit for the first aerial photograph goes to French author and artist Felix Tournachon, who went by the popular name of Nadar. He captured the first aerial photo from a balloon tethered over the Bievre Valley, in 1858. (The oldest aerial photograph in existence is a view of Boston by James Wallace Black, in 1860.) So as far as actual evidence goes, the first verifiable instances of photo reconnaissance work were during World War I. (Early cameras used glass negatives, where the photographer had to change plates between exposures.) Obviously, by the way, choice A is out; Louis Daguerre didn't even begin creating images until the 1830s.

Answer to Question #35: none of the above. You might expect that you would have to have accomplished the thing that your invention was ostensibly designed to do, before you could dash off to the Patent Office! But actually, the Wright brothers submitted their application for a patent in early 1903, nine months *before* their first flight. (They had applied for a patent in 1902, but they were turned down.) As to when a patent was awarded, that is a different matter. Universally recognized as having made one of the most important contributions, namely the identification of an effective method of lateral control (the idea finally protected by their patent) it wasn't actually issued until issued May 22, 1906. Patent number 821,393 was issued on May 22, 1906 to "O. & W. Wright" for a "flying machine".

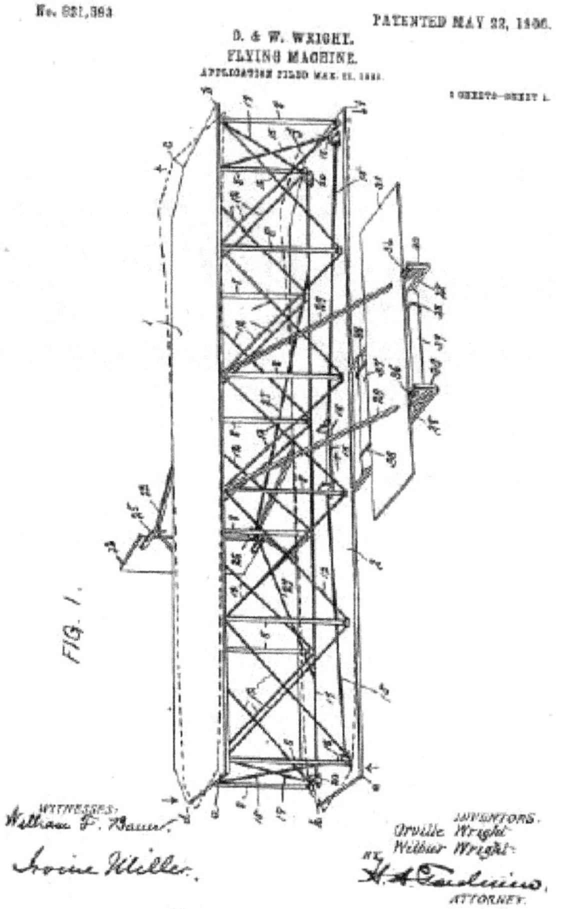

photograph from U.S. Patent Office

Answer to Question #36: It's choice D. Short Mayo who? Rocket scientists weren't the first to try increasing their vehicle's range by adding additional stages. (This *is* a trick question, yes.) Back in 1938, the Short S21 Maia flying boat was wedded to an S20 Mercury seaplane on a flight from Ireland, during which the upper aircraft was separated and flew on to Montreal, Canada with a cargo of mail and newspapers. This was continued, on a route between Southhampton to Alexandria, until the start of WWII. One may wonder which airplane was the first to be built primarily of composite material (at least as far as its exterior surfaces), and the Windecker does seem to be the likely choice. Many airplanes from as far back as the 1930's used at least a small amount of material that was eligible for the term "composite" or at least "synthetic". (In the case of the Windecker, no, it was not carbon fiber, but it *was* the first civil aircraft having an outer structure made almost entirely of fiberglass.)

Answer to Question #37: choice D: Note that the term is and always *has* been "aircraft carrier" and not "air*plane* carrier". And yes, 1861. The aircraft in question here were balloons. The flamboyant showman Thaddeus Lowe, Chief Aeronaut of the Army of the Potomac, oversaw the conversion of the coal barge G. W. Parke Custis for the purpose of tethering, transporting, and launching one or more of the eventual inventory of seven observation balloons used for reconnaissance and artillery direction for the Union Army under General McClellan. Entering service in November of 1861, the Custis frequently towed these balloons on the Potomac River to observe the activities of Confederate forces, until the Army Balloon Corps was disbanded in the middle of 1863.

As I have already said, what was really the "first" aircraft carrier, at least for airplanes, will be argued back and forth, probably for years to come. Was it the Birmingham? the HMS Ark Royal in 1914, or the HMS Furious, later on during WW I? or the Langley, or the Saratoga? Hey, who cares? One thing will remain indisputable though: considering aircraft in general, it has to be 1861!

Answer to Question #38: The right answer is choice A. Oh ye of little faith! Yes, it happened on September 8, 1910, over Wiener-Neustadt, Austria. The brothers Warchalovsky lived to fly again another day (as did one of their passengers, Archduke Leopold-Salvator), although one of them did suffer a broken leg. (And by the way, the first mid-air collision between airliners on scheduled flights occurred in 1922.) Choice C *was* fiction, except for the fact that there really was a Goodyear dirigible acquired by the US Navy, and the collision in choice B did occur and was most likely the first intentional mid-air. The event in choice D was, sadly, also real.

Answer to Question #39: choice C: Sadly, choices A and B are not fiction. Lt. Gavotti's action from choice D doesn't count as a civil action because at the time, Italy and Turkey were at war. This was however one of the first (if not *the* first) dropping of bombs from a military airplane. Going even further back of course, man's inhumanity to man had taken to the skies in balloons over a century earlier. According to the Smithsonian's "Book of Flight", the first recorded use of balloons in warfare was by France, at the Battle of Fleurus, in Belgium, in 1794.

Answer to Question #40: Yes, choice E. We know that Leonardo da Vinci was a genius unlike almost any other known to history. This great Italian painter, sculptor, architect, engineer, musician, and natural philosopher carefully studied the flight of birds and established some fundamental principles, but his preternatural insights did not include even the more intuitive precursors to fluid mechanics, atmospheric physics, or aeronautical engineering. Furthermore, so much written knowledge has been lost to the multiple evils of war, ignorance, and the ravages of time that even credentialed historians don't really know for certain what unsung intellects of even greater proportions may have discovered--perhaps before there were means to document it, and before they vanished from human memory and record. What is known is that the English philosopher, Franciscan monk, and scientist Roger Bacon was interested in the biological and physical sciences. Many discoveries have been credited to him, and he foresaw the extensive use of mechanical cars, boats, and planes. In his book "De mirabili potestate artis et naturae" ("The Wonderful Power of Art and Nature") written ca. 1250 A.D. (but which was not published until much later), he describes a lighter-than-air flying boat, with wings driven by a hand-turned crank mechanism. Bacon followed the maxim "Cease to be ruled by dogmas and authorities; look at the world!" His central goal, it seems, was to prove that during the age of alchemy in which he lived, the "inferiority and indignity of magical power, compared with nature and science." He speculated that, since hot air rose, it could be contained, such that a vehicle could float in the air and move about without touching the ground. (Francesco de Lana, 1631 - 1687, the subject of an earlier trivia question, may in fact have gotten this idea from Roger Bacon.)

Abu'l Qasim Abbas ibn Firnas of Islamic Spain supposedly invented, constructed, and tested a flying machine in about 875 A.D. It has been claimed that Roger Bacon learned of flying machines in fact from Arabic references to Ibn Firnas' machine. (If so, it would antedate Bacon by 500 years and Da Vinci by about some 700 years.) ibn Firnas, who died in the year 888, came to Cordoba to teach music, but became interested in the mechanics of flight. He reportedly constructed a glider and attempted (successfully or otherwise) to fly. Of additional interest but totally irrelevant, a mechanic by the name of Charley Furnas became the world's first airplane passenger (and one of few to fly with both Wilbur and Orville Wright) on May 14, 1908.

One of the first devices to successfully employ the principles essential to rocket flight was a wooden bird. The writings of Aulus Gellius, a Roman, tell a story of a Greek named Archytas who lived in the city of Tarentum, now a part of southern Italy. Somewhere around the year 400 B.C., Archytas mystified and amused the citizens of Tarentum by flying a pigeon made of wood. Escaping steam propelled the bird suspended on wires. The pigeon used the action-reaction principle, which was not stated as a scientific law until the 17th century. About three hundred years after the pigeon, another Greek, Hero of Alexandria, invented a similar rocket-like device called an aeolipile. It, too, used steam as a propulsive gas.

And finally, one must consider what discoveries there may very well have been further to the East. Ancient China, credited with the kite as early as the Fifth

Century BC, as well as the rocket before the time of Roger Bacon, may have synthesized these two inventions somehow, although historical evidence has not thus far revealed any quantitative understanding.

woodcut style image of Roger Bacon, ca. 1250 A.D.;
Abu'l Qasim Abbas ibn Firnas of Islamic Spain, ca. 875 A.D.;
and a Greek experimenter named Archytas, around the year 400 B.C.

Answer to Question #41: Right answer: choice C. His inductive reasoning was flawed because he never ventured beyond speeds above those which exist on what we now know as the "back side" of the power curve! (In addition, there *were* some scale effects for which he did not correctly account.) This speed is one which we know today as either the speed for maximum endurance or, *without* power, the speed for a minimum sink glide.

Samuel Pierpont Langley was actually a professor of physics and astronomy at the University of Pittsburgh. In 1886, the year before he was named to be the assistant secretary of the Smithsonian Institution (originally the National Museum) he went to Buffalo, New York for a meeting of the American Association for the Advancement of Science. His interest was captivated, so the history books say, by the lectures that he attended on the subject of flight. (The Secretary of the museum died several months after Langley's appointment, and Langley then succeeded him as Secretary.)

Then, as now, getting research funded was always an issue, and when he did secure funds, he began to measure how much power was required to lift a weight with a moving wing, using a technique described 50 years earlier by Sir George Cayley. Using a "whirling table" with a wing attached to a 30-foot arm that rotated at up to 70 miles per hour, he measured lift and drag forces. In his 1891 paper, "Experiments in Aerodynamics," he concluded that the higher the speed, the lower the drag. This incorrect conclusion was accepted at the time and named "Langley's Law".

Answer to Question #42: The answer is D. The Wright Flyer was not displayed at the Smithsonian until 1948. Heck, it wasn't even in the *United States* for about 40 years! Up until 1948, the Smithsonian actually displayed Langley's Aerodrome as the first airplane "capable of flight". (The whereabouts of the Flyer are a bit more complicated.) This may have had something to do with the fact that Samuel Pierpont Langley had been the Secretary of the Smithsonian at the time of the Wright Brothers' historic flight, that he had been in feverish competition with them to achieve the first manned, sustained, and powered heavier-than-air flight, *and* that he had been soundly beaten by them (despite all of the financial resources at his disposal). The story is also a bit complicated, but as to the reason for Langley's failure, the biggest reason was probably that his models simply didn't account for the much greater influence of all things affecting a full scale machine (such as for example, drag), or other aerodynamic factors. Langley was Secretary of the Smithsonian (or as it was known until 1911, the United States National Museum) from 1887 to 1906, the year of his death. There was pretty much a fued--and that's the right word for it--going on between Orville Wright and the Smithsonian, from about 1914 to 1943. Although the Flyer was not offered to the Smithsonian until 1910, officials at the museum were still determined to "redeem" the reputation of its former director. Langley's Aerodrome was left with the Smithsonian, which in 1914 asked Glenn Curtiss (whom the Wrights had sued for patent infringement) to rebuild it, and to conduct further tests. With many improvements, he succeeded in getting the Aerodrome airborne for short straight-line flights at Hammondsport, New York, after which it was returned to the Smithsonian (restored to its original unsuccessful configuration), and put on display in 1918 (with the above-mentioned claim that was at best, inaccurate). This indignity was probably what most angered Orville, and which was what precipitated his lending the Flyer to the London Science Museum in 1928.

So where *was* the Flyer during the years before it was ceremoniously unveiled in the Arts and Industries Building of the Smithsonian, on December 17, 1948? Well, in London, mostly, but from the end of 1903 through 1913, the airplane slept in crates inside a shed behind the Wright Cycle Company in Dayton, Ohio. The Miami River flooded in March of 1913, and the Flyer lay covered under mud and water for eleven days. In 1916, Orville and a Wright Company mechanic uncrated the Flyer and restored it for a brief exhibit at MIT, and it was also exhibited at the New York Aero Show in 1917. In 1921, it was again reassembled to provide evidence for a lawsuit against Orville Wright that was filed by the survivors of a California glider builder. The Flyer was also exhibited at the 1924 National Air Races held in Dayton. During this time, many interested American museums approached Orville about acquiring the Flyer. He had considered transferring it to the Science Museum in London as early as 1924, and the construction of its new building in 1928 was a likely reason for his decision to do so. Between 1928 and 1943, Orville negotiated with Smithsonian officials, and went so far as to seek the assistance of Supreme Court Justice (and former president) Taft. But it wasn't until President Roosevelt intervened in 1943 that Orville was persuaded to return the Flyer home. At FDR's request, Orville wrote to the director of the London Science Museum in December 1943, requesting its return.

The Aerodrome continued to be displayed in the Smithsonian's Arts and Industries building (with a revised label) until 1971. It was then removed from public exhibition. And the Wright Flyer? The estate of Orville Wright made sure that the airplane would always be displayed with the following label:

> THE ORIGINAL WRIGHT BROTHERS AEROPLANE THE WORLD'S FIRST POWER-DRIVEN, HEAVIER-THAN-AIR MACHINE IN WHICH MAN MADE FREE, CONTROLLED AND SUSTAINED FLIGHT INVENTED AND BUILT BY WILBUR AND ORVILLE WRIGHT FLOWN BY THEM AT KITTY HAWK, NORTH CAROLINA DECEMBER 17, 1903 BY ORIGINAL SCIENTIFIC RESEARCH THE WRIGHT BROTHERS DISCOVERED THE PRINCIPLES OF HUMAN FLIGHT AS INVENTORS, BUILDERS AND FLYERS THEY FURTHER DEVELOPED THE AEROPLANE TAUGHT MAN TO FLY, AND OPENED THE ERA OF AVIATION.

By the way, you can now see a reproduction of the 1903 Wright Flyer and the *original* Langley Aerodrome, on display together for the first time ever, at the Virginia Air and Space Center, in Hampton, VA.

Answer to Question #43: B. on June 10, 1965, at London's Heathrow airport. (1929 *would* be a tad early, but that's when Jimmy Doolittle made history's first take-off and landing by the use of instruments alone.) An ILS Category IIIc is an approach procedure which allows a landing without a decision height minimum and without a runway visual range minimum. The motivation behind this event was that BEA and other European airlines were losing revenue due to frequent weather days, so BEA decided to develop an automatic landing system. Prior to this, autopilot systems worked only in cruise. Autoland could fly down to 12 feet on a CAT-IIIb landing without the pilot touching the controls. The first such passenger flight was in June of 1965. Choice C represents the year of the first test landing of the Space Shuttle Enterprise. (The first landing of STS-1, after a return from orbit, was in 1981.)

Answer to Question #44: It is choice C. It wasn't the Montgolfier brothers. But it was 74 years earlier and, ahem...indoors. In early August of the year 1709, a Brazilian Jesuit priest, naturalist, and mathematician named Bartolmeu Lourenco de Gusmao gave a demonstration flight of a small envelope of heated air in Lisbon at the court of King Johan V of Portugal. As with its eventual descendants, the source of lift for the small bag was heat from a flame, and things were going quite well, until they realized that the royal draperies were going to join in the fun, too.

Answer to Question #45: Choice D is right. At first one might have seen choice B and guessed that Yeager's flight was so secret that they didn't even take a picture. (Well, I don't believe they did, actually, at least not while he was supersonic.) Then despite the fact that the question uses the word "until", you might have seen choice C and figured that perhaps the Germans had used a high speed camera to film a test rocket or something. Nope. It's even more devious than that. The first photographed supersonic flight wasn't taken to commemorate an airplane, or a rocket. It was to record the flight of a bullet, and it was published by none other than Ernst Mach himself, in 1887.

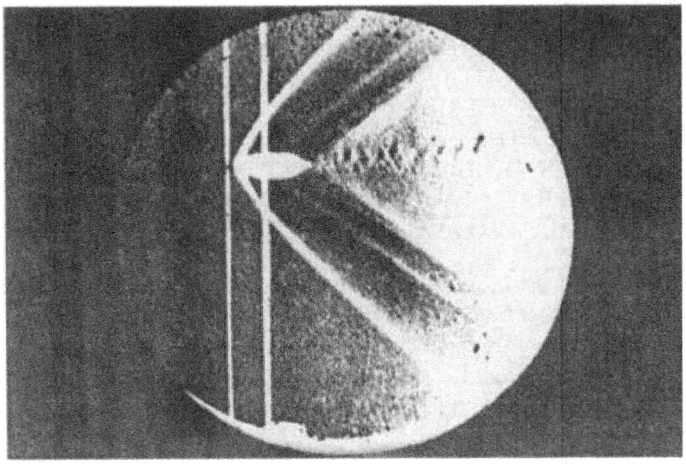

first photograph of a supersonic flight, taken of a bullet in 1887 by Ernst Mach

Answer to Question #46: Choice C. The international Q-codes for radio telegraphy were instituted in 1912, at the Radiotelegraph Convention, held in London. Back then of course, radio communication was almost exclusively via continuous wave ("CW"), meaning Morse code, and there was a significant advantage in being able to transmit a great deal of information in very little time. The QAA to QNZ series, where the second letter can be from A to N--and there are about 180 of them--are reserved for the aeronautical service. (If you are curious, a useful reference is the ICAO Procedures for Air Navigation Services, Abbreviations and Codes, Document 8400.) Many ITU Q codes are still in frequent use today, especially in the amateur radio community, and even though they may be spoken out loud, they still convey information. For example to "QSY" means to switch to another frequency. But two of the aviation Q codes in particular continue to have significance in the aviation community. These are QNH, which signifies an indicated height above sea level, and QFE, where the altimeter indicates height above the airport. QNH of course is the altimeter setting provided in ATIS information, and by ATC. For example, if you were at the Denver/Stapleton airport, and the altimeter setting was 29.92, using QNH, it would show about 5,430 feet, but using QFE, it would be zero. The QFE type of setting is actually used by many pilots (such as those practicing aerobatics). It actually would make good safety sense during instrument approaches; rather than having to remember some weird-ass number for a decision height, wouldn't it be easier to simply look out for 200 feet? In Europe, QFE gets wider coverage. (Choice A, by the way, mentions maritime services. They also have their own range of Q codes, ranging from QO to QQ.)

There is one teeny problem associated with the practical translation of altimeter settings. Unlike the United States, where altimeter settings are reported in inches of mercury, the international usage is hectoPascals, or millibars (which mean the same thing, number-wise: 1 mb = 1 hPA). A US pilot landing in Europe who forgets this might get an altimeter setting of "992", thinking it means 29.92 (which is equal to about 1013.2 hPA), and set his altimeter accordingly. However, *they* would *really* have meant 992 hPa, which is equivalent to about 29.29 inches of mercury, and which is 630 feet lower! Hmmm...as Desmond Llewelyn might have said, bit of a sticky wicket, 007...

Answer to Question #47: choice B. Four *HUNDRED*? Good golly, we probably had quite a crop of candidates for the laughing academy, but not quite *that* many. Believe it or not, starting in 1799, there were just over 200 aviation-related patents filed *before* the patent (number 821,393) that was granted to Orville and Wilbur Wright on May 22, 1906. To their credit, many were ostensibly sound and sober, though there were a few misguided crackpots out there in left field. About 175 were actually awarded and had patent numbers assigned, and about an equal number were for actual flying machines, as opposed to means of propulsion, components, etc. (Incidentally, we had our first patent law in 1790, *before* we had thirteen states; the first patent statute was passed by the Congress of the twelve United States about seven weeks before Rhode Island ratified the Constitution.) Among the titles for the various types of flying machines were such things as "aerial steam car", "balloon locomotives", "apparatus for navigating the air", "flying ship", "flying machine" (Otto Lilienthal, 1895), "soaring machine" (Octave Chanute, 1897), "navigable balloon" (Graf Ferdinand Zeppelin, 1899), "aeroplane" (the first such named, by one Silas Conyne, in 1902), and "aerial vehicle" (Alexander Graham Bell, 1904). There were many patents that were titled "flying machine" (as was that of the Wright Brothers) or "flying-machines", and which were awarded *before* that of the Wright Brothers (about a dozen and a half, actually). And there were many other similar titles, now somewhat archaic, which one would expect of the 19th and early 20th centuries. In addition, there were well over 100 *more* aviation patents filed *after* that of the Wright Brothers, even before 1910. (There was also another patent filed by "O. & W. Wright", in 1909, titled "Mechanism For Flexing the Rudder Of A Flying-Machine Or the Like".)

Answer to Question #48: It's choice A. Glenn Curtiss also held the first US pilot's license, and built the world's first practical seaplane (which did *not* have wheels) as well as the first amphibious airplane (that *did* have wheels, which also retracted), among many other firsts in aviation and other fields as well. White Wing, also known as Drome #2, had motorcycle wheels and tires, as well as a steerable nose wheel (and ailerons)! He made his first flight in this aircraft on his 30th birthday: May 21, 1908. He is credited with over 500 inventions, but he didn't patent a great number of them, preferring to share knowledge and advance the science of flight. He is also credited with the first use of ailerons, although this idea may have come from Alexander Graham Bell, or another individual with whom Curtiss worked in Bell's Aerial Experiment Association beginning in October 1907. (Actually, the very earliest credit goes to one M.P.W. Boulton, a British inventor who patented them in 1868, and their first use was probably by the French aviation pioneer Robert Esnault-Pelterie, in 1904.) Curtiss' contributions to aviation included tricycle landing gear, enclosed cockpits, the hand-grip throttle (for motorcycles), and many other new ideas. Most surprising is a quote from Orville Wright, in a reply to the publisher of *Scientific American*, that pneumatic wheels did not seem to belong on an airplane: "Personally, I think the flying machines of the future will start from tracks, or from [a] special apparatus."

Answer to Question #49: It's C. Although there is some controversy regarding the answer to who invented the first thermometer, Galileo Galilei is generally credited with its invention. He built an early form of this instrument in about 1593. The word thermometer originates from the Greek word "therme" which means heat, and "meter", which implies a measurement over some kind of scale. It consisted of a glass flask, about the size of a small egg, with a neck about 16 inches long and as fine as a straw of wheat. He warmed the flask with his hands, then turned its mouth upside down into a vessel placed underneath, in which there was a little water. When he took his hands from the flask, the decreasing temperature of the inverted vessel resulted in the contraction of the air within it, which in turn changed the level of the liquid partly filling the vessel's long neck: the water began to rise, and mounted to somewhat over eight inches above the level of the water in the vessel. Galileo never published anything stating that he divided the thermometer into measurements of equal degrees, but in a letter dated 1638, another scientist, Benedetto Castelli, describes Galileo's experiment and implies that Galileo did indeed use a scale when examining "the degrees of heat and cold." However, in the same way that our altimeters do not account for temperature, his thermometer was inaccurate because of the reverse: it ignored the effect of atmospheric pressure (although air pressure was not known about at the time). At the start of the seventeenth century, there was no way to quantify heat, except perhaps in a very rough way as when a physician put his hand on a patient's forehead and diagnosed "fever heat." Another Italian inventor, Santorio Santorio, and Galileo's friend Gianfrancesco Sagredo (both in Venice) were the first inventors to put a numerical scale on the instrument (by marking reference temperatures with threads tied around the glass tube). Santorio published a reference to his thermometer in 1612 (which at the time was known as a thermoscope). It had thus become a full-fledged air thermometer. The first series of quantitative meteorological observations date from this period. In other parts of Europe the inventor Cornelis Drebbel and Robert Fludd developed similar instruments. Their nearly simultaneous (and probably independent) discoveries illustrate the seventeenth-century trend toward quantification of natural phenomena--an essential part of the "mathematization" of nature. The liquid-in-glass thermometer was developed in the 1630s, but an accepted standard of temperature remained elusive.

In the early eighteenth century, universal temperature scales based on several fiduciary points (e.g. a mixture of ice and brine, a mixture of ice and water, body temperature, the boiling point of water) were developed by Daniel Gabriel Fahrenheit (1686-1736), Anders Celsius (1701-1744), and René-Antoine Ferchault de Réaumur (1683-1757). Of these, the first two are still in use, and the system of Celsius (extended to become an absolute scale in the nineteenth century) has become the standard scientific temperature scale. Fahrenheit was the German physicist who invented the alcohol thermometer in 1709, and the mercury thermometer in 1714. (Mercury is a better choice because it does not freeze like water and contracts and expands on a smaller scale than does water because it is more dense.) In 1724, he introduced the temperature scale that bears his name: the Fahrenheit Scale. The Celsius scale, devised in 1742 by Swedish Astronomer Anders Celsius, has 100 degrees between the freezing point (0 C) and boiling point (100 C) of pure water at sea level pressure. (He was also responsible for building Sweden's Uppsala Observatory in 1741, and was its first Director.) The Fahrenheit scale is the English unit and the Celsius scale is the metric unit, used first by the scientific community. A simple formula can be used to convert from the English to the metric or vice-versa. $C = 5/9 (F - 32)$ *or* $F = 9/5 (C) + 32$. The term "Celsius" was adopted in 1948 by an international conference on weights and measures. In 1848, the British physicist Sir William Thomson, Baron Kelvin of Largs, Lord Kelvin of Scotland (1824 - 1907), took the whole process one step further with his development of the Kelvin scale by proposing a system that used the degree Celsius but was keyed to absolute zero (- 273.15°C). The Kelvin scale represents the last word in thermal scales measuring the ultimate extremes of hot and cold. Among his other achievements was the 1852 discovery of the "Joule-Thomson Effect" of gases, his work on the first transatlantic telegraph cable (for which he was knighted), as well as his inventing of the mirror galvanometer used in cable signaling, the siphon recorder, the mechanical tide predictor, and an improved ship's compass. The absolute scale related to the Fahrenheit degree is known as the Rankine (R°) scale, named after the Scottish engineer and physicist William John Macquorn Rankine, who proposed it in 1859.

Answer to Question #50: It's choice C. The first airplane to land on a ship using jet power was a Ryan FR-1 Fireball compound fighter. It was powered by a front-mounted Wright R-1820-72W radial engine, and in the rear, a GE I-16 turbojet. (The Navy had its doubts that jet power was suitable for carrier operations.) When Ensign Jake West headed in towards the USS Wake Island on the 6th of November, 1945, his Wright radial quit during the approach, so he continued under jet power only. The first pure "jet only" aircraft to land on a carrier was less than a year later, on July 21 1946. It was a McDonnell-Douglas XFD-1 Phantom (choice A, and which was a prototype) onto the USS Franklin D. Roosevelt. (Production aircraft had an "FH-1" designator.) And the world's first pure jet to *operate from* an aircraft carrier was the de Havilland Vampire (choice B), first landed on the HMS Ocean on by Lt. Cmdr. E. M. Brown on December 3, 1945. Choice D, the Heinkel He-178, was the world's first jet airplane, but it didn't land on any ship!

Answer to Question #51: It's choice A: on March 28, 1910 to be exact. With very little flying experience, the French marine engineer Henri Fabre succeeded in taking off from the Gulf of Fos at Martigues (the Mediterranean, on the French Riviera near Marseilles) in his canard seaplane, called a hydravion. It was also known as "The Duck" (*en Français*, Le Canard). It had three pontoon-like floats, looked somewhat like a dragonfly in reverse, and was powered by a 50 horsepower Gnome engine. On that date, it flew about 1650 feet. I assume he landed as well, although how gracefully, I don't know. (About two months later, it was wrecked when it nose-dived too steeply during a "landing".) The American Glenn Curtiss is credited with the world's first practical seaplane, on January 26, 1911 (and which he called a hydroplane). It was more of a "flying boat" with a single hull, which he tested in San Diego Bay. TOM Sopwith (a.k.a. Thomas Octave Murdoch Sopwith, 1888-1989, a.k.a. Sir Thomas) also made an early contribution to the science of flying from and back onto water with a design which he called the "Bat Boat". (And yes, you read that right; he lived to the age of 101.) The Wright Model CH was a Model C "hydroplane".

Answer to Question #52: It's D. During the early years of aeronautics, when flight made the transition from being an art, then to an industry, and finally to a science, airplane construction took many forms. From the awkward kites of the earliest years, to trussed structures of a more streamlined design, form slowly evolved towards function as engineering principles from other disciplines were adapted to the construction of airplanes. In this type of construction, the fuselage of the plane consisted of a structure strong enough to bear the stresses of flight, reducing the need for external bracing. The monocoque design produced a streamlined aircraft that was lighter and created less drag. Glenn Curtiss did use a plywood monocoque construction in his 1918 Model CB, and Igor Sikorsky's 1913 S-9 had a shell-like structure, also. The stressed metal skin stretched over a rib and spar framework, as in the DC-3, was popular in the 1930s. John Northrup's Vega series of airplanes in the 1920's were noted for their streamlined design. But the man who started it all was the young designer for Deperdussin, Louis Bechereau. My guess is that he simply did what many an inventor does: he saw a potential application of parallel knowledge from another discipline, in this case perhaps, structural engineering.

For starters, two things contribute to a structure's strength: its material, and its shape. A given material can be be strong in compression, but weak in tension, yet due to its shape it can still offer protective value. In two dimensions, an outstanding example of this is the arch. In three dimensions, bird eggs and igloos are two examples. The beak of a feeble chick can easily break its tension-weak shell (mostly calcium carbonate, with a proteinaceous cuticle) from the inside, yet if you tried to crush an egg by pushing in on opposite ends, you would need a force hundreds of times greater in order to do it. The trick to making a structure strong with a tension-weak material is to shape the structure so forces create compression and avoid tension. Concrete is comparatively tension-weak, yet as we can see anywhere we look, strong in compression. Metal, even a thin layer of it, offers the same properties. Bechereau's 1911 design helped usher in aviation's Belle Epoque, won the Gordon Bennett Trophy in 1912, and became the first airplane to exceed 100 mph. (It also won again, in 1913).

Answer to Question #53: choice B. In a way, mother nature herself came up with the idea first. A number of plant seeds take this form in order to expedite a wider dispersal throughout their environment. The two papery, translucent membranous wings of the tropical liana seed, or the seeds of the tropical Asian climbing gourd (below) and others may have impressed early aviation pioneers. Although it is believed that Jack Northrup was the "father of the flying wing" (the XB-35 or its predecessor, the N-1M) in reality, he really wasn't. Over three decades earlier, other designers had the same idea. One example of this was Hugo Junkers, in about 1910. Other examples of that era who are less familiar, such as René Arnoux and John William Dunne, also designed and flew flying wing airplanes. Even somewhat earlier were Igo Etrich and Franz Wels. More strikingly familiar were the 1930s designs of the brothers Walter and Reimar Horten. Following is a montage of illustrations from nature and early flying wing designs, adapted from archival images. Included are images of the tropical Asian climbing gourd, the XB-35 and its predecessor, the N-1M, early designs of Hugo Junkers, René Arnoux, John William Dunne, Igo Etrich, and Franz Wels, 1930s designs of the brothers Walter and Reimar Horten, and the modern B-2.

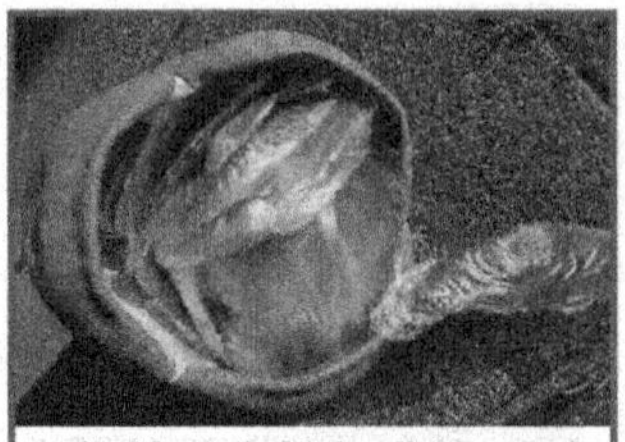

seeds of the tropical Asian climbing gourd

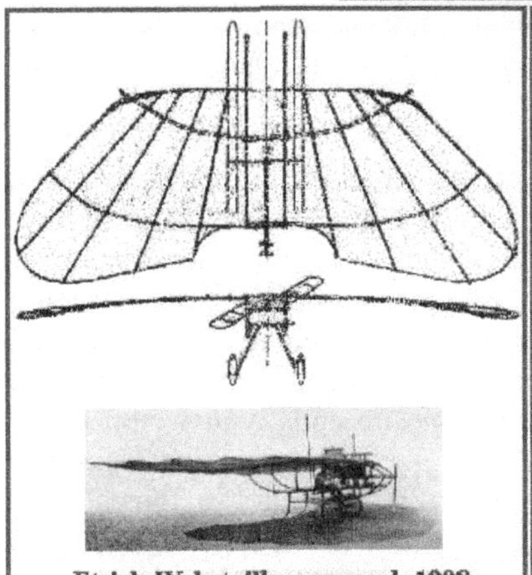

Etrich-Wels tailless parasol, 1908

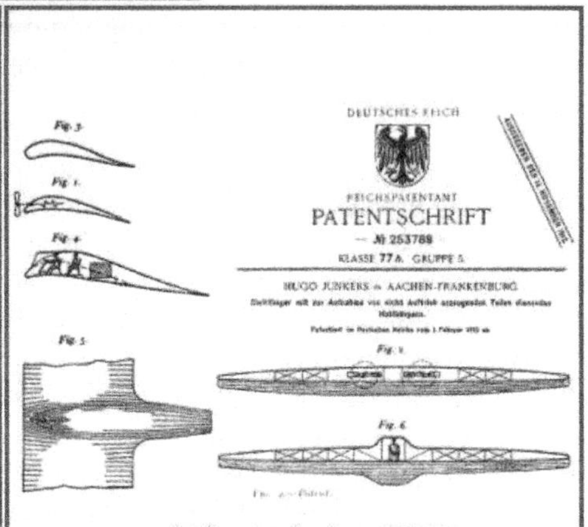

Junkers patent, ca. 1910

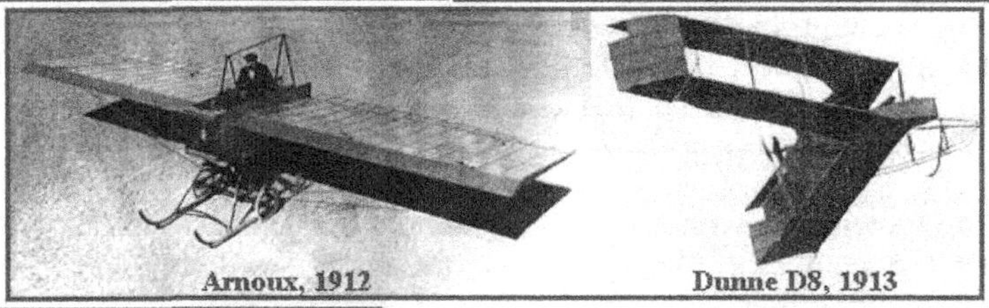

Arnoux, 1912

Dunne D8, 1913

Horten Go-229 (Ho IX), ca. 1944

N-1M, 1940 → Northrup XB-35

B-2

Answer to Question #54: In terms of what went out the factory door, as opposed to what you might encounter on the flight line, the answer is (disappointing as such a lack of an industry agreement may be) that although aircraft manufactured after 1976 in the U.S. and Canada often *do* have airspeed indicators with primary markings based on indicated airspeed in knots, there is apparently no regulatory policy or guidance to which there is universal adherence. Why is this something that is disappointing? Because of the fact that, like assembly lines, things work best and get accepted, understood, and followed most widely when they're standardized. It's harder to hop in and out of different model years of the same Skyhawk for example, and have to remember entirely different units for best rate of climb, best glide speed, etc. And these differences are all the more significant since a knot is 15% more than is a mile per hour. So although for those with experience flying many different aircraft, and for whom choice C may have *seemed* like it was the best answer, in reality, it isn't. It's choice B. First, a little background, before we get to the facts and further explanations:

The knot of course is one nautical mile per hour, which is a bit over 15% greater than one statute mile per hour. Before ships had instruments with which to measure their progress through the water, speed was measured by heaving a rope overboard with a log attached. By measuring the length of the rope played out over time as the ship pulled away, they determined their approximate speed. But after a while, they tied regularly spaced knots, and simply counted those. Formal recognition of this unit however took awhile. The Joint Army and Navy Board on Aeronautics, approved by both the Secretaries of War and the Navy in 1919 (and before that in 1916, as the Joint Army and Navy Board on Aeronautic Cognizance), and then in 1920 shortened to just the Aeronautical Board, agreed unanimously only on June 26, 1946 that the knot and the nautical mile be adopted by the then Army Air Forces and Navy as standard aeronautical units of speed and distance, and directed that use of the terms be specified "in all future procurement of air speed indicators, charts, related equipment and future issues of applicable handbooks and technical orders." Of course, it had been in common use long before that; this was more a formality than anything else. In the area of civil aviation though, it hasn't been mandated. We usually see older aircraft having airspeed indicator markings for calibrated airspeed, in statute miles per hour. (And to this day, somewhere around 90% of US ultralights use airspeed indicators that read in miles per hour.) Generally, it is true that many US and Canadian aircraft manufactured after 1976 have their airspeed indicators with primary markings based on indicated airspeed, in knots. The "FARs" say, in Sections 1323(A) of Part 23 (for normal, utility, acrobatic, and commuter airplanes), Part 25 (for transport category airplanes), Part 27 (for normal category rotorcraft), and Part 29 (for transport category rotorcraft) that airspeed instruments must be "calibrated to indicate true airspeed" but it doesn't say "knots." In the subsequent paragraphs in those same sections, the regulations give the maximum allowed tolerances for system errors for those calibrations as being the greater of a certain calibrated airspeed percentage or a *certain number of knots* (three percent or five knots, for airplanes and normal category rotorcraft). But this does not constitute a binding requirement as to primary indicating units, only a reference. There is no obligation for a manufacturer to maintain strict adherence to the implication of these FAA rules, nor is there any similar Joint Aviation Authority (JAA) ruling, nor has there been any NPRM in the Federal Register. You also won't find standards in FAA Technical Standard Order C2d on airspeed instruments (and if you pay just $59 to the kindly folks at the Society of Automotive Engineers, they'll let you look at Aerospace Standard 8019, "Airspeed Instruments" to which this TSO defers. You also will not see anything in Advisory Circular 23.1309-1C, "Equipment, Systems, and Installations in Part 23 Airplanes." All that said, the non-regulatory status of what you'll see on an airspeed indicator has been confirmed for me by both Cessna's technical services support folks in Wichita, as well as AOPA's technical specialists, who helped me with research on this question. The bottom line seems to be, you'd better pay attention to the units as well as the numbers!

Okay, so we know a knot is a nautical mile per hour, but then, what is a nautical mile? Although it has long been a useful fact that each geographic minute of latitude is equivalent to about one nautical mile, it is not precisely correct. As has been mentioned in other Trivia questions, the Earth isn't a perfect sphere, but is instead flatter at the poles (by about 23 out of some 6880 nmi of diameter). The result is that a minute of arc is about 6046 feet at the equator and 6109 feet at the poles, the current international value being exactly 1852 meters (or 6076.1155 feet). The meter itself was originally defined in terms of the Earth, as one 10 millionth of the distance from the North Pole to the Equator (at the meridian of Paris). The original measurement is now known to be incorrect, but it is now nonetheless enshrined in platinum rods and sanctified in terms of cesium wavelengths.

(I know: whoopee.)

Answer to Question #55: Uh...well, all four. Some purists might say choice A, some might say D. One could conceivably trace its roots back to the Balloon Corps of the Army of the Potomac, or 1863 when Congress authorized the Civil War Signal Corps. Primarily concerned with visual signaling, the Signal Corps also supplied telephone and telegraph communications, began the use of telephones in combat, used combat photography, and of course, balloons. But the precursor to the US Air Force originated on August 1, 1907 as the Aeronautical Division of the Army Signal Corps (later called the Aviation Section of the US Signal Corps). Total personnel strength was three: one Captain, one Corporal, and one Private First Class (who later went AWOL). In July 1914, the Aviation Section of the Signal Corps had 30 aircraft and 40 pilots. By 1916, total personnel strength was 311, and by 1918, it was about 195,000. The Signal Corps was the branch of the US Army that took charge of "all matters pertaining to military ballooning, air machines, and all kindred subjects." However, after America entered World War I in April of 1917, it was evident that the Signal Corps wasn't the right organization for our combat aviators, and the War Department created the Army Air Service on May 24, 1918. The forerunner of the US Army Air Corps, the Army Air Service was separated from the Signal Corps and made a separate combatant arm of the US Army in 1920, and renamed to the US Army

Air Corps on July 2, 1926. A revised, expanded (and again renamed) Army Air Force replaced it on June 20, 1941. Although the Lampert Committee of the House of Represetatives had first proposed an Air Force independent of the Army and Navy back in 1925, the National Security Act which made it official wasn't until September 18, 1947.

Answer to Question #56: A. The ambiguity inherent in precisely quantifying the numbers of aerial victories in the confusing and (to say the least) stressful situation of aerial combat, where simply staying alive had to take precedence over keeping an accurate tally of who shot whom, and when, must be taken into account. Quantifying numbers of downed aircraft often bordered on the indiscriminate, and during the First World War many pilots flying for America originally came from the French Air Service or the Royal Flying Corps. (The same applied to a few of our WWII aces from the UK.) During the Korean War, about half of the "aces" (using the definition of five or more victories) had also scored during prior service in WWII. (So if you count everything, the number of US aces in Korea could be 56, or if you only considered US aerial combat that took place during the conflict, it might be only 41.) During Vietnam, Randy Cunningham and Richard Ritchie are credited with five kills, while Robin Olds had "only" four. But he still had *thirteen* victories from his service during World War II, two decades before! Does that mean he was no longer an "ace"? (I don't think that would be entirely fair, do you?) So the point is that such numbers will always be approximate. However, one general observation emerges from these numbers, vague though they are: During WWI, the few top American aces had scores of about twenty. During World War II, there were far fewer (less than two percent) credited with that many or more, and during Korea, the top aces could claim somewhere about 15. And to put things in perspective , Captain Charles D. DeBellevue, the leading US Air Force combat ace in Vietnam, shot down six enemy planes.

Fairly comprehensive listings of aces from all wars, and all countries, can be found on many web sites. One example is http://www.au.af.mil/au/awc/awcgate/aces/aces.htm. based on these data, the approximate numbers and victories of all American aces would look something like this:

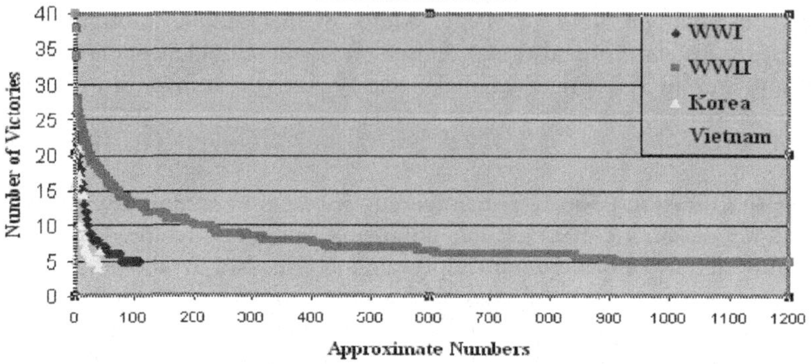

Ignoring for the moment the trailing thousand of so from WWII (those having less than a dozen victories) allows a closer look at relative performances during the other three wars. Apparently, the greater scale and refined lethality with which war was waged in the air may have been responsible for more "kills" in WWII, but then despite the smaller numbers in the relatively smaller conflicts in Korea and Vietnam, victories got progressively "harder" to come by as each side got better at tactics and defensive measures.

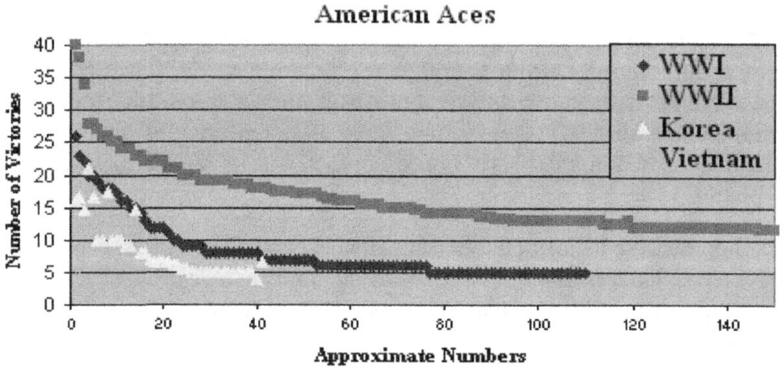

Answer to Question #57: False. There are also rumored incidents involving nuclear weapons belonging to *other* countries. Actually, the term "incident" is a tad misleading, and the word "accident" might be more fair. Applied to ourselves, that would hopefully be the case, because we've had our share of nuclear "oopsies": around 30 of them. Although the release of radiation sometimes occurred (along with the horrendous clean-up logistics), none involved an actual thermonuclear detonation. However...the conventional explosives that are used to *start* an atomic explosion *have* caused the widespread release of fissionable material. The two best known such events were that at Palomares in Spain, in January of 1966 and near Thule in Northern Greenland, also in January, 1968...

In 1966, we dropped *four* thermonuclear bombs on Spain. Yes, Spain. (Each weapon had about a one Megaton yield.) It was Palomares, Spain, to be exact. The term "dropped" here is *also* a bit misleading; it's more like they "fell". (No, we didn't do it on purpose.) In mid-January, 1966, a SAC KC-135 tanker and a B-52 bomber carrying four hydrogen bombs collided while refueling over the Mediterranean coast. Three of them (as well as pieces of the wreckage) fell near Palomares. The fourth bomb fell into the Mediterranean and was later recovered, intact. Although one bomb landed in a dry creek bed, the other two of those three bombs exploded during their fall (the conventional "TNT" part, I hasten to add, and not the plutonium), and fissile material was scattered over about a one square mile area. About 1,400 tons of contaminated material were removed for storage back in the USA. The over 700 people who were in the vicinity at the time have been evaluated over the subsequent quarter-century, but even the Defense Nuclear Agency said, in 1975, that the total extent of the damage will never be known.

Four seems to be an unlucky number here. Just over two years later, a B-52 with four one-Megaton weapons sustained an in-flight fire and crashed on the ice about 10 miles from Thule Air Base in Greenland. The impact with the ice detonated the weapons' conventional explosives, spreading plutonium-239 over a small area (this time, only several acres in size). Several pounds of plutonium were found on the ice, as well as on many thousand pieces of wreckage. An unknown amount was also carried out to sea with the smoke that rose into the air. About a pound of Pu-239 (broken up into very small pieces) was subsequently found on the seabed. Immediately after this second accident, American aircraft were forbidden to carry such weapons on routine flights.

And lest you think we only did it to other countries, we've also done it to ourselves about equally as much. Before either of these incidents for example, in February 1958, a B-47 bomber had a midair with an F-86 fighter southeast of Savannah, Georgia. The pilot was ordered to jettison the bomb in order to land safely. Unfortunately, it was nighttime and no one paid enough attention to exactly where it fell: somewhere along the southern shore of uninhabited Little Tybee Island. It has never been found. A few weeks later incidentally, another H-bomb was accidentally dropped, this time near Florence, South Carolina.

Answer to Question #58: B. In late November, 1913, the first known aerial combat took place over Naco, Mexico, between two Americans: Phil Rader, flying for Gen. Victoriano Huerta, and Dean Ivan Lamb, with Venustiano Carranza (two opposing leaders during the Mexican Revolution). Although the exact details are unknown, about a dozen pistol shots were exchanged. Neither pilot was wounded, and neither aircraft sustained damage.

Answer to Question #59: Many factors contribute to the safety rating of an airline and its accident history, such as maintenance, management, operating procedures, training, the age of its fleet, and routes flown. In addition, accidents are often caused by circumstances beyond an airline's control, including navigational aids, instrumentation at airports, ATC errors, other aircraft at fault in collisions, sabotage (and terrorists), design flaws, and also of course, weather. Even if all factors were considered, an airline accident is still a very rare event, although of course poor weather information, crew miscalculations, and a few questionable flying procedures have taken their toll. Still, the probability of your being killed on a single flight is approximately eight million to one. To put it another way, if an otherwise immortal passenger boarded a flight at random, once per day, *every* day, it would be at least 20,000 *years* before he or she would die in an airline accident.

El Al, Israel's national airline, and Qantas appear to be two of the safest in the world. In the 1988 movie *Rain Main*, Dustin Hoffman stated that Qantas was the world's safest. While at that time it may have been true, and no flying kangaroos have exactly fallen from the sky, on September 23 1999, a Qantas Airlines flight did crash on landing, in Bangkok. (There were no fatalities, however.) To date, Qantas has never lost a passenger...since airliners began flying jets, that is. The last recorded El Al fatality was in 1955, and for Qantas (which stands for Queensland And Northern Territory Aerial Services), it was in 1951. Incidentally Qantas is one of the world's oldest airlines (ca. 1920), with only KLM being older. Also incidentally, KLM stands for Koninklijke Luchtvaartmaatschappij NV, which in Dutch means Royal Air Transportation Company (like you really needed to know that). Before 1951 though, the accident picture wasn't so rosy; Qantas *has* lost some passengers & aircraft in its long history:

de Havilland DH-9C G-AUED 24 Mar 1927 - 3 fatalities
de Havilland DH-86 VH-USG 15 Nov 1934 - 4 fatalities
de Havilland DH-86 VH-USE 20 Feb 1942 - 9 fatalities
Short S-23 (flying boat) VH-ADU 22 Apr 1943 - 13 fatalities
Lockheed 18 Lodestar VH-CAB 26 Nov 1943 - 15 fatalities
Short S-23 (flying boat) VH-ABB 11 Oct 1944 - 1 fatality
Lancastrian VH-EAS 07 April 1949 - no fatalities
de Havilland Drover II VH-EBQ 16 Jul 1951 - 7 fatalities
Lockheed L1049 VH-EAC 24 August 1960 - no fatalities

On the other hand, a number of less well-known airlines haven't done so poorly, either, and actually, just based on pure numbers (with no regard for passenger volume, nor any attempt to normalize the data), their numbers are even better. As of February of 2003, Hawaiian Airlines hasn't had a fatal accident since 1929; PLUNA, not since 1936; and Arkia Israli Airlines, not since 1950. So, technically, the answer is, for the moment, choice D, although realistically, one might side with Dustin. (AirTran used to be known as ValuJet, and no, there's no letter "u" in Qantas.)

Answer to Question #60: It is true that in 1910, Glen Curtiss dropped dummy bombs on the buoyed outline of a battleship on a lake in upstate New York. And the fact that the Wright Brothers first approached the military with their design in 1905 is fairly well known. Earlier still, while one can't presume to apply the term "national" to these efforts, a Union Army Balloon Corps did exist during the Civil War, for which Thaddeus Lowe bore the grand title of "Chief Aeronaut". However, if you took the subject title seriously, this is a give-away. Thomas Jefferson, third US president and author of the Declaration of Independence, himself presented plans for the use of aviation (balloons, in this case) in the young nation's interest, quite early on. The idea may have come from someone else in correspondence, or from an acquaintance, or perhaps from elsewhere within the new government. But yes, the answer is choice D. His "Aeronavigation Plan" is laying around somewhere in the Library of Congress, but here are some excerpts from some of Thomas Jefferson's "Writings":

HOT-AIR BALLOONS
To Dr. Philip Turpin
Annapolis, Apr. 28, 1784

DEAR SIR-Supposing you may not have received intelligence to be relied on as to the reality & extent of the late discovery of traversing the air in ballons, & having lately perused a book in which everything is brought together on that subject as low down as Decemb. last, I will give you a detail of it. I will state the several experiments, with the most interesting circumstances attending them, by way of table, which will give you a clearer view & in less compass.

They suppose the minimum of these ballons to be of 6 inches diameter: these are constructed of gold-beaters' skin & filled with inflammeable air. this air produced from iron-filings, the vitriolic acid & distilled water is, in weight, to Atmospheric air as 7. to 43. on an average of the trials: & when produced from the filings of Zinc, the Marine acid & distilled water, is to the Atmospheric air as 5. to 53. or 1. to 10 1/2. but Montgolfier's air is half the weight of Atmospheric. this is produced by burning straw & wool. the straw must be dry & open, & the wool shred very fine, so that they may make a clear flame, with as little smoke as possible. 50 lb. of straw & 5 lb. of wool filled the ballons of Oct. 19. & Nov. 21. in five minutes. these ballons contained 60,000 cubic feet. no analysis of this air is given us. Mons'r de Saintford the author of the book, gives us a very great & useless display of Mathematical learning, which certainly has as yet had very little to do with this discovery: & when he comes to the chemical investigations, which are interesting, he sais little. the ballons sometimes were torn by the pressure of the internal air being insufficiently counteracted in the higher regions of the Atmosphere. these rents were of 6. or 7. f. length, yet the machine descended with a gentle equable motion & not with an accelerated one. by the trials at Versailles & Champ de Mars it appears that they will go with a moderate wind 150. leagues in 24 hours. there are yet two principal desiderata. 1. the cheapest & easiest process of making the lightest inflammable air. 2. an envelopment which will be light, strong, impervious to the air & proof against rain. supplies of gas are desireable.

Too, without being obliged to carry fire with the machine: for in those in which men ascended there was a store of straw & wool laid in the gallery which surrounded the bottom of the ballon & in which the men stood, & a chaffing dish of 3. feet cube in which they burnt the materials to supply air. it is conjectured that these machines may be guided by oars & raised & depressed by having vessels wherein, by the aid of pumps, they can produce a vacuum or condensation of atmospheric air at will. they are, from some new circumstances, strengthened in the opinion that there are generally opposite or different currents in the atmosphere: & that if the current next the earth is not in the direction which suits you, by ascending higher you may find one that does. between these there is probably a region of eddy where you may be stationary if philosophical experiments be your object. the uses of this discovery are suggested to be 1. transportation of commodities under some circumstances. 2. traversing deserts, countries possessed by an enemy, or ravaged by infectious disorders, pathless & inaccessible mountains. 3. conveying intelligence into a beseiged place, or perhaps enterprising on it, reconnoitring an army &c. 4. throwing new lights on the thermometer, barometer, hygrometer, rain, snow, hail, wind & other phenomena of which the Atmosphere is the theatre. 5. the discovery of the pole which is but one day's journey in a baloon. from where the ice has hitherto stopped adventurers. 6. raising weights; lightening ships over bars. 7. housebreaking, smuggling &c. some of these objects are ludicrous, others serious, important & probable. I will give you the figures of the baloons on the last page.

Congress has determined to adjourn on the 3d of June to meet in November at Trenton. a vessel arrived here yesterday which left London the 25th of March. she brings papers to the 20th of that month. mr. Pitt was still in place, supported by the city of London, the nation in general, & the House of Lords. still however the majority in the H. of commons was against him, tho reduced to 12. it was thought the parliament would be dissolved.

Be so good as to present my dutiful respects to my uncle & aunt & to be assured of the esteem with which I am Dr. Sir

your friend & serv't

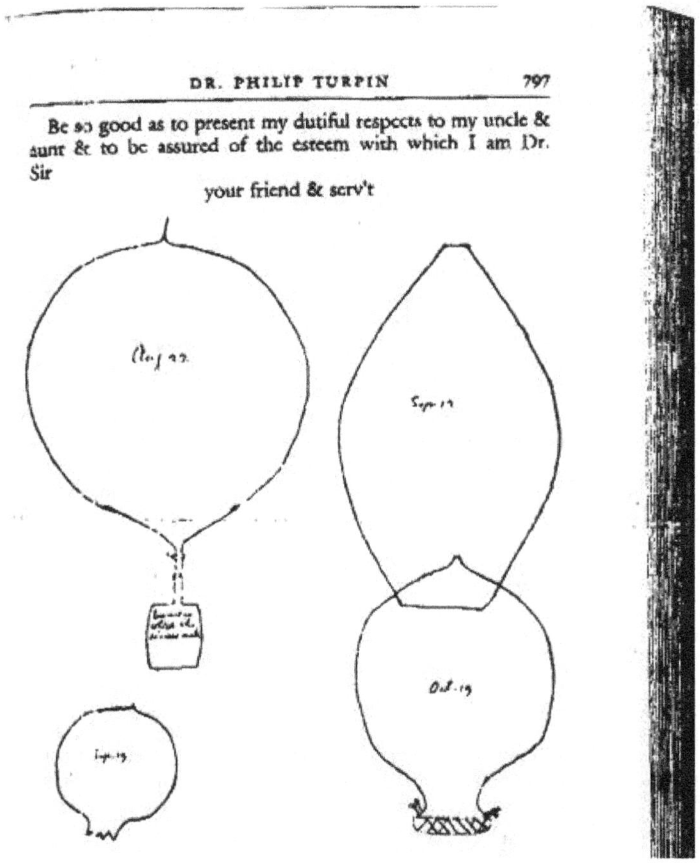

DR. PHILIP TURPIN 797

Be so good as to present my dutiful respects to my uncle & aunt & to be assured of the esteem with which I am Dr. Sir

your friend & serv't

Date of experiment 1783	Place of experiment	Height of Ballon in feet	Diameter of Ballon in feet	Materials of which the Envelop was made	Materials from which the internal air was produced	Weight comp'd with common air	Weight the Ballon could raise	Horizontal distance to which it went	Ascent in the air	
June 5	Annonay		36	paper	straw & wool burnt	1/2	lbs. 490	1200 toises	1000 toises	
Aug. 27	Champ de Mars		12	taffety gum elastic	iron filings unindicated	1/6	35	5 leagues	unknown	
Sept. 12	Academie Koiale	70	...	canvas and paper	straw & wool	1/2	1250		...	
Sept. 19	Versailles	57	41	cloth	straw & wool	1/2	696	1700 toises	300 toises	a sheep, duck &cock aboard
Oct. 19	Paris	70	46	...	straw & wool	1/2	...	50 toises	54 toises	2 men aboard
Nov. 21	Chateau de	70	46	...	...	...	1700	5000 toises in 20'	1600 toises	2 men aboard
Dec.	...	...	...	...	...	...	...	twenty odd miles	1600 toises	2 men ab'd barom. fell from 28 to 180-Faren hi; thermom. fell 36 deg.

Answer to Question #61: We've all seen something like this on military aircraft, and if you built model airplanes as a kid, you probably remember applying insignia decals to the wings or fuselage. American aircraft have displayed several different national insignia since the days of World War I. The very first one was simply a five-pointed red star, painted on JN-3 "Jenny" biplanes of the U.S. Signal Corps, back in 1916. The US adopted its first official national aircraft insignia in 1917. It was a white star with a central red disc, on a blue circular background. (US aircraft in Europe during WWI also used three concentric circles of white, black, and red, similar to the national insignia of Britain and France and differing only in their color sequence). Several months after Pearl Harbor though, the red disc was *removed* from the center of the white star to avoid any association with Japan's "rising sun". In June of 1943 the US added wing bars, which improved visibility, and in January 1947, horizontal red bars were added to the white rectangles on either side of the circle. Usually displayed on the fuselage, top left wing, and bottom right wing, this remains the standard national symbol for US military aircraft flying today. The answer is choice C.

Answer to Question #62: As far as careers advanced (or at least of those hoping to lurch forward out of obscurity) it is sometimes not what you know, but whom. One case in point is U.S. Army Lt. George H. Boyle, dashing in appearance, but inept as far as actual airmanship. He was an aspiring air mail pilot, and he enjoyed the conspicuous convenience of being engaged to the daughter of the Interstate Commerce Commissioner who had influenced the government to allow Army pilots to fly the mail. Technically, he was the first airmail pilot to fly along an established route, on May 15, 1918. (Airmail flights had been flown however, as early as 1911.) While President Woodrow Wilson and the Assistant Secretary of the Navy, Franklin D. Roosevelt, looked on, he began his flight to Philadelphia from Washington D.C. by departing to the South (in the wrong direction). Despite having had maps plastered to his legs by his superior, he promptly became lost and landed in a soft field, nosing over. Then the next day (and with a new propeller), that same superior officer gave him explicit instructions for a second attempt, which involved the simple directive that he should keep the Chesapeake Bay on his right. He even flew alongside him up to the northern end of the bay. Once on his own however, he followed his orders to the letter (sorry) and he flew in a big circle, around the bay. When he landed (and didn't actually crash) to take on more fuel, he then actually used some rudimentary navigation and reached the general vicinity of his intended destination. He chose a landing site (but the wrong one, and much too short), and wrecked the plane (only this time, more thoroughly) when he crash-landed on a golf course. Naturally, by that point he had seriously jeopardized his budding airmail career. To be fair, pilots had only a simple magnetic compass and the maps of the time showed large cities, but little of elevations or landmarks. Still, you have to wonder. As regards the above list of airmail pilots, all of the above technically did have that title at one time. But one, choice D in this case, didn't have it for long.

Answer to Question #63: According to "Flying As It Was" by Gerry Casey, the person who actually performed "blind flying" tests before Jimmy Doolittle was one Wesley L Smith. An aspiring opera singer, a degreed engineering student (with honors, no less), as well as a pilot, his first attempts at such involved simply ruddering a compass heading and using the tachometer and engine noise (as "primary" indicators) for altitude control. On May 20, 1920, he used a fixed loop antenna to fly a triangular course while under a hood (only in this case I suspect that it was the "cover your head" variety). During the following nine years, he worked with Howard Salisbury, who (the book says) developed the turn-and-bank indicator as well as a sensitive altimeter, and on February 6, 1929, over seven months before Jimmy Doolittle, he successfully landed a Douglas M-4 somewhere near Cleveland in actual zero-zero conditions (which had developed during his flight) right in the center of the airport. This may be simply dumb luck, but it was done somewhat intentionally (mostly because he *had* to). It's B.

Answer to Question #64: Since the days of Morse code communications, the letter "R" was used to mean, “I have received your message.” It was first used because of its obvious familiarity from the days of early radiotelephony because it represented the letter R in an early (but not the first) phonetic alphabet. Since 1927, the word "Roger" was used (up until WWII, in the US) to represent the letter R. Despite the subsequent switch to "Romeo" for the letter R in the ITU or NATO phonetic alphabet, it stuck. The right answer is choice C.

Answer to Question #65: It's B. This story is schmaltzed up a bit, but here it is. It's mostly true. (I didn't write it, however. Its author is unknown to me.) First, the story, then, the *rest* of the story:

> "During the course of World War II, many people gained fame in one way or another. One man was Butch O'Hare. He was a fighter pilot assigned to an aircraft carrier in the Pacific. One time his entire squadron was assigned to fly a particular mission. After he was airborne, he looked at his fuel gauge and realized that someone had forgotten to fill his fuel tank. Because of this, he would not have enough fuel to complete his mission and get back to his ship. His flight leader told him to leave formation and return.
>
> As he was returning to the mother ship, he could see a squadron of Japanese Zeroes heading toward the fleet to attack. And with all the fighter planes gone, the fleet was almost defenseless. His was the only opportunity to distract and divert them.
>
> Single-handedly, he dove into the formation of Japanese planes and attacked them. The American fighter planes were rigged with cameras, so that as they flew and fought, pictures were taken so pilots could learn more about the terrain, enemy maneuvers, etc. Butch dove at them and shot until all his ammunition was gone, then he would dive and try to clip off a wing or tail or anything that would make the enemy planes unfit to fly. He did anything he could to keep them from reaching the American ships.
>
> Finally, the Japanese squadron took off in another direction, and Butch O'Hare and his fighter, both badly shot up, limped back to the carrier. He told his story, but not until the film from the camera on his plane was developed, did they realize the extent he really went to, to protect his fleet. He was recognized as a hero and given one of the nations highest military honors. O'Hare International Airport in Chicago was named after him.
>
> Prior to this time in Chicago, there was a man named Easy Eddie. He was working for a man you've all heard about, Al Capone. Al Capone wasn't famous for anything heroic, but he was notorious for the murders he'd committed and the illegal things he'd done.
>
> Easy Eddie was Al Capone's lawyer and he was very good. In fact, because of his skill, he was able to keep Al Capone out of jail. To show his appreciation, Al Capone paid him very well. He not only earned big money, he would get extra things, like a residence that filled an entire Chicago city block. The house was fenced, and he had live-in help and all of the conveniences of the day.
>
> Easy Eddie had a son. He loved his son and gave him all the best things while he was growing up; clothes, cars, and a good education. And

> because he loved his son he tried to teach him right from wrong. But one thing he couldn't give his son was a good name, and a good example.
>
> Easy Eddie decided that this was much more important than all the riches he had given him. So, he went to the authorities in order to rectify the wrong he had done. In order to tell the truth, it meant he must testify against Al Capone, and he knew that Al Capone would do his best to have him killed.
>
> But he wanted most of all to try to be an example and to do the best he could to give back to his son, a good name. So he testified. Within the year, he was shot and killed on a lonely street in Chicago. These may sound like two unrelated stories, but Butch O'Hare was Easy Eddie's son."

Well, first of all, Butch got some airplanes all right, but they were "Betty" bombers (or "zippos" as they were referred to by U.S. forces, because they burst into flames so easily), so they weren't as fearsome an opponent as the Japanese Zero.

The rest of the story behind the renaming of Chicago's airport in honor of Butch O'Hare is considerably more involved. In the late 1940's, the city had approached the (still fairly new) U.S. Air Force with a proposal to build a terminal on the south side of Orchard Field (as it was then known) for airline operations, as Midway was becoming crowded. Surprisingly, the Air Force agreed, and the original airline terminal was the result. It was later demolished to make room for the current terminal (the B and C concourses) which is occupied primarily by United Airlines. At some point airline operations started to crowd the Air Force and they told the city to leave. The city's answer was pretty much along the lines of "No, we're here to stay", to which the Air Force's response was, in effect, "No, this is a military facility and you are crowding us; we were here first, so you leave". It became a very bitter battle and the issue of whether the city of Chicago could be evicted from Orchard Field wound up in Federal court. A Federal judge (coincidentally, located in the city of Chicago) ruled that the city couldn't be evicted. At that point, the city (in a blatant effort to rub the Air Force's nose in it) renamed the airport not *only* for a Navy aviator, but one with a known underworld connection! (This is not in any way an attempt to denigrate Butch O'Hare; he was an honorable pilot and a hero, but his selection was intended as an affront to the Air Force in the renaming of the airport.)

Answer to Question #66: As the story goes, a well-to-do entrepreneur from Cincinnati named T. Higbee Embry had a vision of educating the people of his city about the potential of flight, and in the early 1920s, he drove to a local airport to take a flying lesson. A barnstormer by the name of John Paul Riddle took Embry for a ride in a Curtiss Jenny. Seven days later, Embry soloed in a Waco. A close friendship resulted, and on December 17, 1925, exactly 22 years after the historic flight of the Wright Flyer, Riddle and Embry founded the Embry-Riddle

Company at Lunken Airport in Cincinnati, where it was located from 1926 to 1930. They were the first distributorship for Waco. The company soon opened the Embry-Riddle School of Aviation. For a complete story on this "Harvard of the skies" see The Sky is Home, by John McCollister and Diann Davis. Sorry folks, no flying squirrel, and no moose. The answer is choice A.

early photograph of Embry-Riddle Company headquarters, located at Lunken Field in Cincinnati, Ohio, from 1926 to 1930

Answer to Question #67: Lindbergh made the first solo transatlantic flight in 1927, but his crossing of the Atlantic by air was not the first. That was done eight years earlier by U.S. Navy Lieutenant Commander Albert C. Read and his crew of five in May of 1919, flying a Navy-Curtiss flying-boat NC-4 from Newfoundland to Portugal via the Azores, powered by four 400-hp Liberty engines. Although it was the only one of four to complete the trip (which at just over three weeks, was hardly non-stop), it certainly didn't qualify for the £10,000 prize offered by the British newspaper the *Daily Mail*. The rules required a flight in less than 72 hours across the Atlantic by an airplane or airship in either direction between the British Isles and the United States, Canada or Newfoundland. (Newfoundland was not yet part of Canada.) The first real non-stop crossing was by RAF Captain John Alcock and Lieutenant Arthur Whitten Brown, in a Vickers F.B.27 Vimy bomber. They took off near St. Johns, Newfoundland on June 14,1919, and landed June 15 at Clifden in Ireland, sixteen hours, and twenty-seven minutes later. Many others made the crossing, such as the LZ-126 (Luftschiff Zeppelin). The ship, called ZR-3 (for "Zeppelin Rigid") was flown nonstop from Germany to New Jersey in October 1924 by Dr. Hugo Eckener and crew. (Lindbergh did win the $25,000 prize offered by Raymond Orteig for the first flight in an airplane from New York to France.) Depending upon how you count them (not-quite-completed crossings, etc.) and also perhaps who might be considered as essential crewmembers and who would not, Lindbergh was somewhere between the 85th and the 92nd (choice C).

So, then...

- The first through sixth places go to Commander Read and his crew (Breese, Hinton, Rhoads, Rodd, and Stone) of the NC-4 in May of 1919.
- Numbers seven and eight (June, 1919): Alcock and Brown.
- Numbers nine through 42 (July, 1919): Maj. George Scott of the RAF and his crew plus several British VIPs (Angus, Browdie, Burgess, Cooke, Cross, Durrant, Edwards, Evenden, Forteath, Gent, Graham, Gray, Greenland, Harris, Luck, Maitland, Mayes, Mort, Northeast, Parker, Powell, Pritchard, Ripley, Robinson, Scull, Shotter, Smith, Thirlwall, Turner, and Watson), two U.S. observers (Lieutenant Commander Lansdowne of the Navy and Lieutenant Commander Hensley of the Army), and one stowaway (William Ballantyne) flew in the British dirigible R.34 nonstop from Scotland to New York and/or from New York to Scotland. (Ballantyne, Edwards, and Lansdowne made only the westward passage while Angus, Hensley, and Turner made only the eastward trip. All others made the round trip.)
- Numbers 43, 44, 45, and 46: In August of 1924, Lieutenants Lowell Smith, Leslie Arnold, Erik Nelson, and John Harding of the U.S. Army flew in two army Douglas World Cruisers, the Chicago and the New Orleans from England to Labrador, via Iceland and Greenland. This was part of an "around-the-world" flight that these four completed in September 1924. In April and May 1924 they had also been, together with Lieutenant Leigh P. Wade and Staff Sergeant Henry H. Ogden in the Boston, the first to cross the Pacific by air (with stops in the Aleutians).
- Numbers 47 through 78 (October of 1924): Dr. Hugo Eckener and his crew (of 27: Auer, Belser, Christ, Fischer, Fleming, Freund, Grofzinger, Kiefer, Knorr, Ladwig, Lang, Lehmann, Leichtle, Martin, Marx, Pabst, Praff, Pruss, Sammt, Scherz, von Schiller, Schwendt, Siegle, Specy, Tassler, Tielmann, and Wittemann), plus 4 U.S. military observers (Captain Steele, Commander Klein, and Lieutenant Commander Kraus of the Navy and Major Kennedy of the Army) in the dirigible LZ-126, aka ZR-3 (later christened U.S.S. Los Angeles).
- Numbers 79 through 81 (January, 1926): Major Ramón Franco of the Spanish Army and Captain Ruiz de Alda of the Spanish Navy and their mechanic Pablo Rada flew from Spain to Brazil with several stops in the Dornier Wal flying boat Plus Ultra. (Within ten years, Ramón Franco's fame faded, thanks to his brother Francisco, the Spanish dictator.)

- Numbers 82, 83, and 84 (March of 1927) Captain Sarmento de Beires and Captain Jorge de Castilho of the Portuguese Army and their mechanic Manuel Gouveia flew from Portugal to Brazil, with several stops in the Dornier Wal flying boat Argos.

- And at the very earliest, we have number 85 (May 1927): Captain Charles A. Lindbergh of the U.S. Army Air Corps Reserve, nonstop, solo, from New York to Paris in the Ryan monoplane Spirit of St. Louis.

Answer to Question #68: In today's US dollars, that would be about $17,000--or back then, about $850. So either A or C could be counted as being correct.

Answer to Question #69: It's C (for Central). Helen Richey was hired by Central Airlines in 1934 and flew her first trip on December 31 that year, between Washington, DC, and Detroit, in a Ford Tri-Motor. After experiencing blatant discrimination, she resigned within ten months.

Answer to Question #70: The answer is choice (B). It was a 20-cent parcel post stamp, and it was issued on December 16, 1912. It was actually the first stamp in the *world* to depict an airplane (and its arrival was some six years before the first airmail stamp).

Answer to Question #71: In 1826, the will of the English scientist James Smithson left over half a million dollars to the United States for the purpose of establishing an institution "for the increase and diffusion of knowledge." An Act of Congress established it in 1846, with a Board of Regents, and plans for building a museum with geological and mineralogical cabinets, a chemical laboratory, a gallery of art, lecture rooms, and a library. The answer is choice (D): 1876. Now, over a century and a quarter later, the National Air and Space Museum has about 325 aircraft, 30,000 items related to aviation, about 9,000 space-related artifacts, over 4,000 pieces of aviation art, 1,750,000 aviation and space related images, over 40,000 reference volumes, and among other things, a Space Shuttle.

Chapter 2: Greater Perspectives

Even though pilots are fortunate in having their transcendent perch from which to view the world as it passes by beneath, and even though flying affords a greater understanding of what makes the atmosphere (as well as the rest of the planet) such an interesting place to be, still, everything is always relative. In other words, there are always wider horizons. There are certainly opportunities to acquire (and reasons to appreciate) knowledge drawn from other disciplines; it helps us integrate our total understanding, and improves our ability to make useful inferences later on. Here are some sidelong glances by way of just a few examples as to just how a dash of this and that from (in most of these cases) the physical sciences can help us learn, enjoy, and remember.

Subject: Is the earth REALLY flat? Just A little Bit?

Question 1: How "flat" *is* the earth?

A) Actually, friction with the earth's atmosphere over eons of time and the greater moisture at lower latitudes has caused more erosion--the end result of which is that the earth has actually gotten "skinnier", (equatorial diameter less than that between the poles), not flatter and fatter!
B) just one 13,000th: the distance between the poles is about 1600 feet less than the equatorial diameter.
C) Flat as a pancake, yessiree. All this orbit stuff's a big hoax.
D) Even solid rock has some give when 24 sextillion tons are being rotated around at a few hundred knots. The "flattening factor" is about one 297th. To a rough approximation, we are flying around over an oblate spheroid, not just a sphere! If you're a REAL purist, it's more like a pear, or even "the geoid".

Subject: The Longest Day

Question 2: One of the results of how accurately we can now measure time with atomic clocks has been the realization that the rate of the earth's rotation is actually slowing! True, it doesn't slow down a great deal, but it IS slowing down---but by how much?

A) one hour each year
B) one minute a year
C) about a second every year
D) one second a century

Subject: Distant Horizons...

Question 3: Under what circumstances could a glider actually glide to the visual horizon, without the benefit of a tailwind?

A) Never. Even a glider with carbon fiber wings and an aspect ratio of extraterrestrial proportions could not approach a glide ratio of 100 to 1 in earth's atmosphere.
B) Using ridge lift, some sailplane pilots have soared well beyond several horizons.
C) It might, using magneto-hydrodynamics, similar to certain experimental submarines.
D) As one ascends, the ratio of horizon distance to absolute altitude actually decreases, and well under 20,000 feet, this value actually approaches the glide ratio of some sailplanes.

Subject: Time And Tide

Question 4: When landing at an airport near the seashore, why might you care what the tide tables say?

A) Offshore winds often accompany high tides, and there are very often onshore winds during low tide. This can be helpful information for choosing which way to land.
B) Birds are more numerous at low tide. You have a greater chance of avoiding a bird strike at high tide.
C) Lacking a calendar, you can use the highest high (spring) tides to predict the next full moon, when planning a night flight.
D) At low tide, you will have more beach available as an emergency landing site.

Subject: Dance Of the Magnets

Question 5: Why is there a difference between the magnetic variation for an airport, and the VOR located at that same airport?

A) Magnetic variation changes over time and remains constant to within one one-hundredth of a degree for any given location, whereas the radials from a VOR might vary by ten times that over a single day.
B) Because of the obliquity of the ecliptic (the tilt of the earth's axis), there is a very slow inductive coupling between the earth's magnetic field and the solar wind, which causes the earth's magnetic poles to precess about a common point over a cycle of roughly 26,000 years. This affects the earth's magnetic field, and it is the reason that we have to paint new numbers on our runways every few decades. However, if a VOR could last 26,000 years, it wouldn't change one bit.

C) When a navaid is first installed, the antenna is physically oriented to True North. A potentiometer adjustment is then made to slave the navaid with Magnetic North. This action matches the isogonic line making it agree with a magnetic compass. Initially, these values are the same, but the magnetic variation of the earth changes at a rate of about 50.27 seconds of arc per year. Navaids are commissioned and remain online, 24/7. Although periodic maintenance is performed as needed, a re-slaving to match the isogonic value requires a total navaid shut down, re-alignment, and a re-certification flight check. Only when the navaid is out of tolerance by at least +/-6 degrees will a re-slaving procedure be initiated allowing the navaid and airport magnetic variation to match again.

D) Baloney. (the same for choice A, *and* choice B (at least as far as the solar wind and the obliquity jazz) But this precession is real, and because of it, in about 12,000 years, the Northern Hemisphere will have summer in December and winter in June (if we still have months with those names then). But it has nothing to do with the earth's magnetic field. No, the magnetic poles wander mostly because of one thing, which is the "secular" variation due to the chaotic churning and shifting in the earth's molten core. (There's also a secondary artifact due to continental drift.) Yes, it's true about the six degrees, the re-slaving, and the re-certification flight check. Actually, choice C came from the FAA's own Aviation System Standards, and NACO's FAQ page. But they didn't get the story exactly straight. The change in magnetic variation at all locations on earth (or any two chosen at random) is not the same single value. Yes, it's got to do with the earth's molten innards. However, the variation in the earth's magnetic field is *not predictable* (at least yet, and certainly not to the degree of precision which the FAA's number implies). There may be some short-term zonal periodicity, or it may be predictable in one area, more or less, over a few years, but certainly not in *all* locations, indefinitely! The rate of change in other words, is changing, and that is *at any one spot*! Geologists and the like actually have different words for what we think of as magnetic variation. To them, it's magnetic declination. The actual "dip" angle (which results in the compass errors that so plague us) is known as magnetic inclination.

Subject: Round, Round, Round She Goes

Question 6: An airplane departs for a flight around the earth above the equator, at a constant altitude above mean sea level (one sufficiently high enough to preclude being influenced by topography), and in a westerly direction. At the same time, another airplane begins its flight in a direction differing by 90 degrees, and performs a pole-to-pole circumnavigation at an identical altitude above sea level, but along a great circle of constant longitude (first west, then east). If both airplanes start from the same place at the same time, return to their starting point, and have the same climb and descent profile, which one, if any, would fly a measurably greater distance?

A) the one headed pole-ward, due to cumulative crosswind correction
B) The equatorial flight would travel the greatest distance.
C) neither one: Both circumferences are within one tenth of a kilometer of each other.
D) There is not enough information given here to determine the correct answer.

Subject: Coming Or Going

Question 7: Why might a pilot learning to sail become at least momentarily disoriented?

A) because wind speeds are typically 100 times faster than ocean currents.
B) because an airplane can head directly into the wind, but a sailboat cannot
C) because pilots are used to sighting courses from altitude, with oblique perspectives towards a progression of checkpoints, whereas sailors can only sight short distances along the surface
D) because wind direction and current direction are indicated in exactly opposite directions

Greater Perspectives: Answers

Answer to Question #1: The answer, of course, is choice D.

Answer to Question #2: The answer is D. It's mostly due to tidal friction, believe it or not. It's about one hour every 450,000 years. So our brachiating ancestors lived 22, 20, maybe 18 hour days, depending on how far back you go. Also, the moon was a tiny bit closer in then (a few dozen miles), which actually would shorten the days a bit more. Starting in 1972 with Coordinated Universal time, we've been (the astronomical weenie technocrats, that is) inserting (or removing) "leap seconds" to keep with International Atomic Time. (They usually do it, if it's needed, on June 30 or December 31.)

Answer to Question #3: The right answer is choice D. From a yes/no standpoint, choice A is incorrect (though the statement about L/D not coming near 100 is true). Choice B is also quite correct, but not as mathematically elegant, and choice C is nonsense. At about 17,000 feet, the horizon-to-absolute altitude ratio is about 50, and at 25,000 feet, the horizon/height ratio is about 40! (Thinner air actually does nothing to a sailplane's L/D, and some gliders have gotten quite high. The absolute altitude record for sailplanes is currently 50,671 feet.)

Answer to Question #4: choice B. Aside from a storm surge, there is little correlation between wind direction and level of the water at the shoreline. There *is* a correlation between tides and lunar phases of course, so this might be a plausible application of one's knowledge of the physical sciences, but these days, who doesn't have a calendar? Don't even think about landing on a beach at low tide, unless you're either on fire or you know for certain that it will be deserted and without hidden hazards.

Answer to Question #5: It's choice D. The annual rate of change in what *we* know as magnetic variation, is one minute of arc per year.

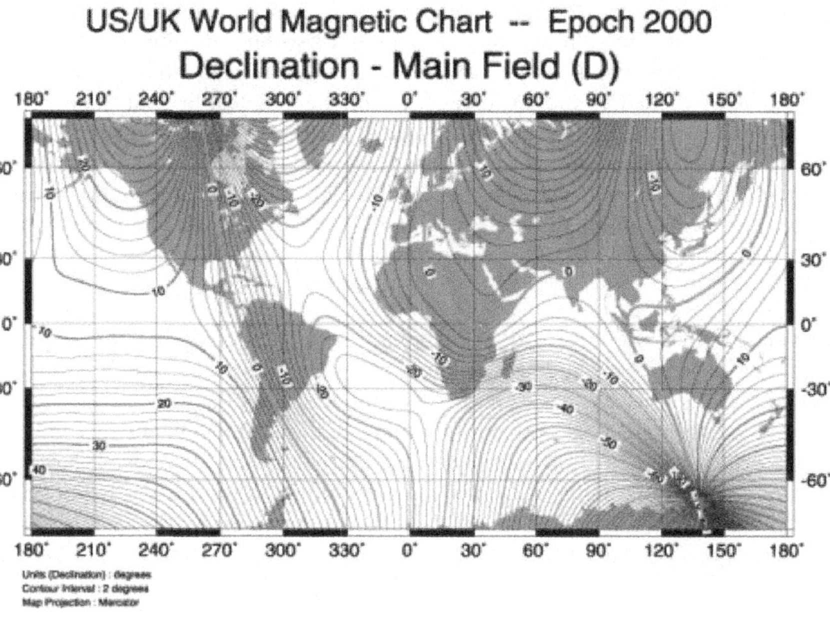

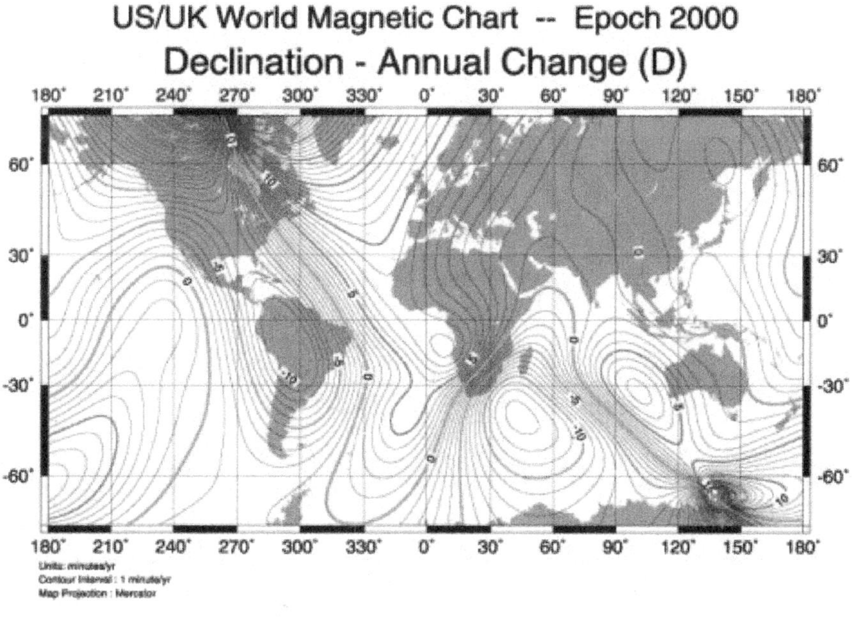

geomagnetic contour interval maps from the USGS
(charts reproduced with permission from U. S. Geological Survey)

Answer to Question #6: choice B. The pole-to-pole flight would be about 36 nautical miles less than the equatorial one. This is due to the fact that the earth isn't a perfect sphere, but (to the first approximation) an oblate spheroid; the earth's equatorial radius is about eleven and a half nautical miles greater than its polar radius. (Actually, the difference would be greatest if you started in the longitudinal vicinity of Denver and headed north or south on your meridional pole-ward circumnavigation, due to the fact that the earth's equatorial cross-section itself isn't even circular, but elliptical. This so-called "triaxiality" however is about two orders of magnitude less than oblateness, and would be very difficult to measure, at least with your airplane.)

Answer to Question #7: As all pilots soon learn, by meteorological convention, winds are depicted using barbs or vectors with directions that are measured clockwise from North, where the "arrow" points *toward* the direction from which the wind is blowing. Thus, a wind direction of 270 degrees means that the winds are westerly, or blowing from the West. However, in the sister science of oceanography (or any other branch of the physical sciences dealing with bodies of water), vector directions denote the direction of mass flow. So in this case, an arrow pointing westward means that the current is flowing from East to West! It's D, all right.

Chapter 3: Airmen, Aircraft, Airports, and Airspace

Here are the stars of the show (us, of course), as well as some of the unusual things that a few of us have flown, done, or the places we've gone, and since pilots often regard the air they fly through as belonging to someone, there are always interesting regulatory quirks regarding our airspace as well. In this chapter, we turn the spotlight on ourselves and the aviation arena.

Subject: Abandon Hope, All Ye Who Enter Here!

Question 1: Apart from temporary flight restrictions, how many ***permanent*** prohibited areas are there in the US?

A) 42
B) 27
C) 9
D) 3

Subject: More On Prohibited Airspace: In Downtown Washington, DC:

Question 2: How old is the block of special use airspace in downtown Washington DC which is known affectionately as P-56?

A) Ancient. It was established by Congress on December 18, 1903.
B) Nah, not THAT old. But it IS old. It's been around since just after World War I, though it has changed shape somewhat. It used to cover just from the White House to the Capitol building, and also included the US Naval Hospital (before it was moved to Bethesda MD).
C) C'mon! It's STILL not THAT old! P-56 (A) was initially set up under an Executive Order in June of 1938. P-56B (the Naval Observatory and the Vice President's residence) was established in May of 1950.
D) It was initially established in January of 1959.

Subject: P-56 Again--Another Angle:

Question 3: About what percent of violations of P-56 are GA aircraft?

A) About half of all recent violations have involved corporate & air carriers.
B) 99 percent of 'em are those gol-dang wild Indian puddle jumpers!
C) The information is classified and has not been released to the public.
D) 20%. Air America has a route over the Mall at 16,000 feet, which is technically below the top of it (which is FL 180). Whenever there's a notice from the CIA, they look the other way.

Subject: Airport Identifier Trivia:

Question 4: What does an "X" tell you about an airport when it's the last character in its three letter "location identifier"?

A) This means that the National Weather Service's original two-letter identifier for airports existing at the time simply had an arbitrary third character appended to it. The "X" probably seemed the most neutral and non-inferential. So what it tells you is that whatever the airfield is, it's probably OLD!
B) This was used to signify the location of beacons that were used to assist early airmail pilots with navigation. So what this means is that the airport was located on an early airmail route. (It's still OLD, but this is actually the correct reason.)
C) The "X" signifies an airport that has, or formerly had, a US Customs office.
D) (A) and (B)

Subject: More Location Identifier Trivia:

Question 5: Location identifiers, once assigned, tend to stick---especially at busy major airports where changing the identifier would mean major headaches. What major metropolitan airport had a change in BOTH its identifier AND its name? (And that's a hint.)

A) This has never been done.
B) Miami changed its identifier from 2FL6 in 1947.
C) Washington National was known as WAS up until 1958, when it was switched to DCA.
D) In December 1963, Idlewild Airport changed its name (and identifier) to JFK.

Subject: Methuselah's Wings:

Question 6: Who is (or was) the world's oldest pilot?

A) Ebenezer Prendersnarf of Concord, New Hampshire is 99 and still an active CFI.
B) There are actually two. Both are 98, although they do not fly alone. Their licenses require that they each fly with a safety pilot.
C) Jeanne Calment, who lived to age 122, not only was (for a time) the world's oldest person, but retained her French sailplane license until the age of 109.
D) Gimme a break! No, you lose a little finesse when you get THAT old! The world's oldest active pilot (so far, anyway) was John Miller of Poughkeepsie NY, at age 102.

Subject: Runway Numbers: It's NOT Cast In Concrete!

Question 7: Under what circumstances might a runway number misrepresent its true (OK, real) magnetic direction?

A) Almost always. Runway numbers are an approximation of their magnetic direction rounded to the nearest multiple of ten degrees. So nine out of ten times, the true magnetic direction will be at least a degree off from what you see painted on the runway, multiplied by ten of course.

B) Choice (A) is true, but there's more! Whenever there are more than three parallel runways, such as at Dallas-Fort. Worth, at least two of them will be deliberately misnumbered. Rules allow for only three sub-designators: L (left), C (center) and R (right). Under these circumstances, they set up two pairs, one of which is "wrong".

C) Choices (A) and (B) are true, but on top of that, the earth's magnetic field is always changing, and the magnetic variation would eventually render those painted numbers obsolete, if you waited long enough. Fortunately, this occurs over a time frame of decades in the US. (Just for the record, if you're up in near the magnetic North Pole in Canada, which is moving towards the northwest at about 10 Km per year, and which has on top of that a diurnal variation, elliptical in shape, of up to 80 Km, you don't use a compass anyway.)

D) Although there are specifications regarding every detail of airport signage and markings, there is actually no rule that says runway numbers must correspond to magnetic direction. If an airport operator wanted to number a runway to avoid the number "13" for example, he or she is entirely free to do so.

Subject: Short Stuff:

Question 8: What (and where) is the world's shortest runway (fixed wing, now) with scheduled airline service? (I don't mean some bush strip in Idaho or a 600 foot mountaintop gash with a 20 degree upslope, like in the 1990 movie Air America!)

A) On American Samoa, there is a strip used by amphibious aircraft operated by Polynesian Airlines which extends just 200 feet into a lagoon. (It used to be longer, but a storm changed the local geography. Now it is just used as a taxiway after a normal water landing. Territorial authorities have not officially changed its status from a runway, however.)

B) Hooey! Creative writing, that's all choice (A) is! The real "shortest runway" is on Tetiaroa, a coral atoll 26 miles North of Tahiti and formerly owned by the actor Marlon Brando. There is scheduled air service from Papeete. The sand runway is 795 feet long. (You stop in a hurry.)

C) On the island of Saba ('Say-bah'), about 28 miles South of St. Maarten and only 5 square miles in size, there is a runway that is about 1300 feet long.
D) none of the above

Subject: The Longest Highway In the Sky.

Question 9: When an instrument pilot files a route, he or she can shorten its description by omitting any navaids (usually VORs) that lie on the same Victor airway. For example, going from Athens, Georgia up to near Richmond, Virginia, even though one passes over several VORs (Greenwood, Charlotte, Liberty, Lawrenceville, Hopewell) and traverses hundreds of miles, you can just enter "AHN V454 HPW" for the route, and the computer "knows" exactly what you mean. So: going from one place to another, what is the LONGEST distance an IFR (or VFR) pilot can travel along the Victor airways (under 18,000 feet) which has the SHORTEST route description?

A) Victor 1 from Jacksonville Florida to Boston, Mass.
B) Victor 3 from Presque Isle Maine, wee wee wee, all the way down to Key West, Florida
C) It's V-16. That airway connects Los Angeles, CA with Boston, MA and is about 2,590 nautical miles long.
D) Victor 10, from Honolulu to San Francisco.

Subject: En Route Altitudes.

Question 10: What is a MOCA, and what is it for?

A) It's airline pilot slang for what the copilot has in his thermos. Maximum Overdrive Caffeine Additive.
B) Minimum Obstruction Clearance Altitude. It is lower than an MEA or Minimum En Route Altitude, often by hundreds of feet, but it does not assure navigation signal reception beyond 22 nm from a facility. Flying at a MOCA, you will clear any obstacles within the lateral limits of the airway, but if you're more than 22 nmi away, you might not know what those are! Therefore, they're for emergency use only.
C) It stands for Military Oceanic Control Area. It is the extension of a MOA out beyond the 12-mile limit offshore. Military training activities are conducted within its boundaries according to international agreements.
D) The Museum of Contemporary Art, in LA (definitely a cool place, but nothing whatsoever to do with aviation though)

Subject: Duh! Which Way Do We Go?

Question 11: Finally! There is now a way to determine left or right traffic at unfamiliar airports, without leafing through your AFD, overflying and looking for a wind sock or tetrahedron, and without bugging folks on the CTAF! It's been staring us in the face for (depending on where you live) several years! What is that?

A) new LARGE TYPE airport signage, and "L" or "R" subscript on untowered airport runways
B) (J. Edgar would *love* this) pending legislation making NO RIGHT TURNS at all airports
C) change in VFR charts during the last several years
D) new AWOS feature

Subject: Reaching Up To Touch Someone

Question 12: What is the tallest man-made object that you might have to watch out for while flying over the North American continent?

A) Including the TV antennas, the top of the Sears Tower is about 1730 feet AGL (above ground level). It's MSL (above mean sea level) height is 2325 feet.
B) Canada's CN tower in Toronto, Ontario is the tallest self-supporting structure, at 1815.5 feet above the ground.
C) The KTHI-TV transmitting tower in Blanchard, ND (northwest of Fargo, ND) is the tallest standing structure at 2,063 feet AGL (3058 MSL).
D) There are actually tethered aerostats along the southern US border which can rise to 15,000 feet.

Subject: Methuselah's Rotors:

Question 13: Who was (or is) the world's oldest helicopter pilot?

A) Frank Robinson, Sr., of Palos Verdes Estates, California, age 88
B) Walter Frumpley, of Zelienople, PA, age 92 (commercial certificate, and still flies for Allegheny Technologies)
C) Morris Lazarus, of Key Biscayne, FL, age 94, also rated and current in single and multi-engine seaplanes
D) John M. Miller of Poughkeepsie, NY, age 102 (He was also a CFI, and yes, the same guy as in question number six...)

Subject: Not A Good Day

Question 14: What unlucky pilot was shot down on May 13, 1917, again on March 13, 1918, and yet again on August 13, 1918?

A) Roland Garros
B) Rene Fonck
C) Lothar von Richthofen
D) Oswald Boelcke

Subject: Taking General Aviation To the Top

Question 15: The first person to fly alone to the true North Pole in a light aircraft was

A) Richard Byrd, 1926
B) Shiela Scott, 1971
C) Gustavus McLeod, 2000
D) Roald Amundsen, 1926

Subject: Just Hanging Out

Question 16: What is the world's record for the longest single hang-gliding flight?

A) eight hours, in 1994. It was also the longest, at 307 miles, between Rock Springs, Wyoming and Stoneham, Colorado.
B) nine and a half hours, Torrey Pines, CA, 1988
C) 34 hours and three minutes, 1986, Oahu
D) two days, three hours, and 12 minutes; Bonavista, Newfoundland, Canada

Subject: Faith and Begorrah!

Question 17: On Friday, August 5, 1938, a million ecstatic New Yorkers greeted Douglas Corrigan on lower Broadway with a parade and a "tickertape snowstorm." (Later on he got a similar reception in Brooklyn.) The headline of the New York Post on that date actually read as follows:

NAGIRROC YAW GNORW LIAH

The sub-heading was:

"MILLION CHEER MADLY IN WELCOME TO FLYER")

What was the reason for this?

A) He had just completed the first solo round-trip flight across the Atlantic
B) He was supposed to fly to Ireland, but somehow wound up in California.
C) He was supposed to have flown to California, but because of a backwards compass, he landed in Ireland instead.
D) Having been denied permission to fly the Atlantic, he did it anyway, with the excuse that he "just got mixed up in the clouds".

Subject: On a Bicycle Built For Two...

Question 18: How many times did Orville and Wilbur Wright fly together?

A) never: Their long-standing policy was never to fly together, so that in case of an accident at least one would remain to carry on their work.
B) once
C) They really had such an agreeement, but they frequently broke it.
D) There was no such rule. Many of their flights after 1909 were dual!

Subject: Lots Of Room To Slow Down

Question 19: The world's longest runway is located at

A) Edwards Air Force Base
B) KSC, Florida
C) Groom Lake, Nevada
D) JFK, in New York
E) Chicago O'Hare

Subject: Looney Tunes

Question 20: What famous Warner Brothers cartoon character has been immortalized on an actual instrument approach procedure? For extra credit, which one is that?

A) Elmer Fudd
B) Woody Woodpecker
C) Tweety Bird
D) Porky Pig

Subject: That Famous Boom

Question 21: In what airplane was the sound barrier supposedly first broken on October 14, 1947?

A) a Lockheed P-80 Shooting Star
B) Glamorous Glennis
C) the Bell X-1
D) the ninth powered flight of the Bell XS-1

Subject: First Pilot's License

Question 22: Who was the first to receive a US pilot's license?

A) Orville Wright
B) Wilbur Wright
C) Glenn Curtiss
D) Bert Lahr

Subject: Just Your Average Runway

Question 23: Based on a recent AOPA eDirectory, what is the average length of public use runways in the United States? The answer is rounded to the nearest foot, but this does not include heliports, almost 14,000 privately owned airports, and some international airports (but it does include seaplane bases).

A) Due to numerous ultralight landing areas, it's only 2484 feet.
B) There are a great number of private (but still public use) strips which are fairly short, and the average length is only 3240 feet.
C) The average runway length in the US is 4289 feet.
D) Because of the growing number of international airports, it's 5279 feet.

Subject: The Last Flight of the Spirit of Saint Louis

Question 24: When did Lindbergh's famous airplane fly for the last time?

A) Never. It's always flying (though in a hover) at the National Air and Space Museum, as well as in our collective memory.
B) Twang those heart strings, Clarence, but no cigar. The real last time was made by Lindbergh himself, in 1954, the same year he won a Pulitzer Prize for his autobiography having the title of that famous airplane, during a book tour. He strongly protested the idea, but acceded to his publicist, albeit with a great deal of reluctance.
C) On April 30, 1928, about three weeks shy of a year after having landed in Paris on his historic flight, Charles Lindbergh landed the plane at Bolling Field, then located on the banks of the Potomac River in Washington DC, where it was dismantled and taken to the Smithsonian's Arts and Industries Building, eleven days later.
D) It was last "flown" by actor James Stewart (the late movie pilot Frank Tallman, actually) in 1957 for the making of the movie of the same name.

Subject: The One and Only

Question 25: There were over 5000 public-use airports in the US, according to a recent AOPA database. Only about 500 of them have a control tower. The vast majority of those airports having a control tower also have more than one runway. A few however, have just one. How few?

A) three
B) five
C) 15
D) 75

Subject: Almost Famous Airplanes

Question 26: Rod Machado is always using poor hapless Two-One-Three-Two-Bravo in his aviation anecdotes. Was there really such an aircraft having that tail number?

A) Yes, but old 32B has long ago been scrapped. He uses that number specifically to avoid referring to any currently certificated aircraft.
B) No, there never was.
C) Yes, it is the same 1949 Luscombe in which he soloed, although it no longer exists.
D) Yes, it is a Luscombe, and 32-Bravo began its "rise to fame" when Rod flew it in about 1972. It still soldiers on, and is now owned by a Texas pilot.

Subject: Il Capo di Tutti i Piloti di Caccia

Question 27: The highest-scoring, most deadly, boss-of-all-bosses, and root'n-toot'n-est fighter pilot ace of all time was

A) Roland Garros
B) Erich Hartmann
C) Manfred von Richthofen
D) Richard Bong

Subject: "Yes, Captain Ma'am!"

Question 28: Who was the first female airline pilot, and in what year was she first employed?

A) Turi Wideroe, 1961
B) Jackie Cochran, 1953
C) Ethyl Mertz, 1951
D) Helen Richey, 1934

Subject: The First American Female Pilot

Question 29: Who was the first US woman pilot?

A) the Baroness Raymonde de la Roche
B) Bessie Coleman
C) Harriet Quimby
D) Katherine Stinson

Subject: The First Woman To Fly Across the Atlantic

Question 30: Who was the first woman to fly across the Atlantic?

A) Anne Morrow Lindbergh
B) Beryl Markham
C) Amelia Earhart
D) Jackie Cochran

Subject: Speed Zone

Question 31: We have supersonic jet fighters, but are they (or any other type of aircraft capable of supersonic flight) ever flown faster than Mach 1 over the continental United States? More than one may apply.

A) Military aircraft flying combat air patrols are authorized to exceed Mach 1 on an as-needed basis.
B) Regular supersonic flight over populated areas was discontinued in the 1960s.
C) There are designated supersonic corridors and other areas in different locations around (and outside of) the Continental U.S. (such as in central Florida, where *double* sonic booms were often heard from returning Space Shuttles)
D) Sonic booms cannot be heard when the aircraft is above FL 600

Subject: Unusual Airplanes

Question 32: What is unusual about N60094?

A) It is currently the only registered experimental aircraft not fitting into any known category (the most commonly used term being "flying saucer").
B) It is the oldest airworthy aircraft in the United States.
C) It is the only private aircraft in the United States that ever transported a US President while in office.
D) It was last flown by Elvis Presley.

Subject: The World's Most Southerly Runway

Question 33: The world's most southerly runway is located in

A) Christchurch, New Zealand (NZCH)
B) Antarctica (NZSP)
C) Capetown, South Africa (FACT)
D) Guardiamarina Zanartu, Chile (SCGZ)
E) Punta Arenas, Chile (SCCI)

Subject: The Longer Runway

Question 34: What is the significance of a runway that is 8069 feet in length?

A) aside from the fact that this is the minimum short field landing distance of a B-747-400 at max gross weight, none at all
B) That is the cutoff point for a different runway symbol on all aeronautical charts.
C) This is the maximum length for a single runway to be depicted on a sectional chart inside a circle that is just under three-sixteenths of an inch wide.
D) When NOAA standards for symbology were redefined in 1965, this was the minimum length for the top five percent of runway lengths in the US, and this determined how large the aerodrome symbol would be.

Subject: The Highest Runway In the East

Question 35: Where and how high is the highest runway east of the Mississippi?

A) Bangda, Georgia, at 4784 feet MSL
B) Burnsville, North Carolina, at 4,432 feet
C) Mitchell Field, North Carolina, at 6684 feet
D) Albuquerque, Tennessee, 6492 feet

Subject: On Beyond Noisy

Question 36: There was an aircraft that, while it may not have had the noise output of the Valkyrie supersonic bomber from another trivia question, produced sound which proved to be even more troublesome than that which puts one's hearing at risk. Which one was it?

A) the Hiller YH-32 Hornet
B) the Bachem Ba-349 Natter
C) the Republic XF-84H Thundershriek
D) the Brewster F2A Buffalo

Subject: 4F

Question 37: In September 1993, the FAA adopted the International Civil Aviation Organization (ICAO) definition of airspace segments. At that time, American pilots learned that the ICAO airspace classifications are named A through G. Mysteriously though, most of us were never told just what Class F was supposed to mean, and just why the US decided not to use it. And for those inquiring minds who wanted to know, there wasn't exactly an avalanche of information on the subject. (There still isn't.) So...just what in blazes *is* Class F, anyway?

A) Outside the US, it is simply any special use airspace.
B) any uncontrolled airspace in which VFR night flight is not allowed

C) in *most* (but not all) countries outside the US, uncontrolled airspace above FL 600
D) uncontrolled airspace in which separation is provided to participating IFR traffic

Subject: The Two Jackies

Question 38: Jacqueline Cochran had humble beginnings on (as she called it) "sawdust road", but eventually she succeeded in life after training as a beautician, and founded a large cosmetics company. She became interested in aviation in large part due to the business motivation of being able to cover more territory, but went on to accomplish many "firsts" for female aviators. She was the winner of the 1938 Bendix Transcontinental Air Race, she was one of the developers of the oxygen mask, and most notably, was the first woman to fly faster than the speed of sound in 1953, in an F-86 Sabre. (She did receive some assistance from a friend of hers who had also shared less than auspicious origins, one named Chuck Yeager.) She also organized and led the Women's Airforce Service Pilots, during WWII. However, there was another Jackie with whom she had a long-standing, although informal, rivalry, each one continually trying to out-do the other in the setting of new aviation records. Who was that?

A) Major Jacquelyn S. Parker
B) Jacqueline Auriol
C) Jaclyn Katharine Wright
D) Jacqueline Lee Bouvier

Subject: The Fighter Without A Gun

Question 39: What fighter airplane had an "F" type designator, but yet carries no guns?

A) the Brewster F2A-3 Buffalo
B) the Grumman F-13
C) the F-117A
D) the Fokker F-10

Subject: Curious Names of Remarkable People

Question 40: Much has been written about Glenn Hammond Curtiss, and he has been mentioned in several of these trivia questions. You might recall reading about the Aerial Experiment Association, founded by Alexander Graham Bell (and which included Glenn Curtiss among its members, along with US Army officer Thomas Selfridge). A curious coincidence is that the chosen site of the AEA was Hammondsport NY, where Glenn Curtiss was located, and (as some may have noticed) the *middle name* of Glenn Curtiss was Hammond. So, was there any connection?

A) Glenn Curtiss' great-grandfather, Hammond Curtiss, was an early settler in the area, and the town was named in his honor.
B) absolutely none: It was only a coincidence.
C) Hammondsport was his mother's home town, and she chose it for no other particular reason.
D) Glenn Curtiss was actually given no middle name. Convinced that a middle name would add some level of credibility to his work, he added it in 1906.

Subject: A New Era In Commuting

Question 41: The first person to use an airplane for transportation on a regular basis (in connection with a non-aviation related occupation) was:

A) William Frederick ("Buffalo Bill") Cody
B) Samuel Franklin Cody
C) Harry Houdini
D) Elmer Sperry

Subject: Stubby

Question 42: What propeller-driven aircraft had a fuselage that was nearly as tall as it was long?

A) The McDonnell XF-85 "Goblin"
B) The Italian Stipa monoplane
C) The Lockheed-Martin GZ-20 and GZ-22
D) The Aero Spacelines 377SG

Subject: Right Through the Ceiling

Question 43: Sailplane pilots ride thermals to gain free altitude. At some point, when the moisture in the air reaches the dew point, it condenses to form a cloud. Assuming that instability continues, such a column of rising air will keep going up, regardless of the fact that it's no longer doing so under visual meteorological conditions (or VMC) as far as the pilot is concerned, although it will do so at a somewhat slower rate, as a rule. (If this were not so, clouds would have little or no vertical dimension.) The question is this: In order to gain as much altitude as possible, is it legal for a sailplane pilot to continue soaring upward in IMC, inside a cloud?

A) Although it is done outside of the United States in Class F airspace, it is illegal in the US.
B) It can be done, but there are some restrictive qualifications with which both pilot and aircraft must be in compliance.

C) Although it used to be perfectly legal prior to the mid-1970s and it is *still* not technically illegal according to the letter of the law, practically speaking, it is no longer approved by ATC.
D) The pilot must have a glider instrument rating, and the aircraft must be equipped for IFR flight.

Subject: The Birds

Question 44: While en route at 4000 feet, ATC issues you the following advisory: "Flock of small birds, twelve o'clock, six miles, last reported at four thousand." What does this signify?

A) It means that you may be crossing paths with an F-117A. The radar returns of such stealth aircraft are often diffracted by the networked mosaic of Center ARSR radar antennas into patterns which resemble a flock of birds.
B) The controller is probably a trainee, and has made an error. Most birds don't show up on radar.
C) You might want to request a higher altitude. Some types of radar can detect flocks of birds, but they may also have been reported by another pilot.
D) Real birds don't show up on radar, but phantom ones can. The shadows could be an artifact of ionospheric refraction, and subsequent diffraction of returns from distant high terrain.

Subject: High Diddle-Diddle, Right Down the Middle

Question 45: What are "Q" Routes?

A) VFR routes which are allowed to traverse Restricted Areas
B) high-altitude routes in Europe and Asia
C) Area Navigation (RNAV) routes
D) low-altitude routes at high *latitudes* which are not "straight line" but which instead have standardized arcs

Subject: It's Not Easy Being Green

Question 46: You take a fuel sample during your preflight, and the color of the sample is green. This means

A) Either the fuel farm has some real vintage fuel left sloshing around, or you've entered a time warp: Grade 100/130 fuel is no longer produced
B) Antifreeze has leaked into your fuel tank.
C) Two fuel grades have been inadvertently mixed, and you've been mis-fueled.
D) It's March 17.

Subject: Class A airspace

Question 47: True or False: It is illegal to fly VFR at or above 18,000 feet MSL.

Subject: Early Curmudgeonry

Question 48: The first recorded noise complaint (one that was aviation related, that is) was in what year?

A) 1949
B) 1974
C) 1907
D) 1928

Subject: The first CFI

Question 49: The world's first (civilian) flight instructor was

A) Orville Wright
B) Frederick W. (Casey) Baldwin
C) Glenn Curtiss
D) Walter Brookins

Subject: Both Sides Of the Mike

Question 50: How many active air traffic controllers in the US are also active pilots?

A) only about one in twenty
B) about one in ten
C) about one in four
D) just over half

Airmen, Aircraft, Airports, and Airspace: Answers

Answer to Question #1: It's choice C. There are only nine parcels of prohibited airspace in the entire US—three of which are in the Washington DC area (which is not surprising). These are P-40 (Camp David); P-56 (A and B); and P-73 (Mount Vernon). The others are P-47 in Amarillo Texas; P-67 in Kennebunkport Maine; P-204, P-205, & P-206 in northern Minnesota; and the newest one, P-49 (effective May 17, 2001), over the Crawford, Texas residence of former President Bush.

Answer to Question #2: The correct choice would be C. (That's my final answer.) The evolution of this particularly contentious but critical parcel of carefully guarded airspace has an interesting history, which was very patiently pieced together with the invaluable help of one Ned Preston, former FAA Historian, now retired. Here it is, in one picture, which is an adaptation using an FAA aeronautical chart, annotated by the author, which shows the history of Prohibited Area P-56:

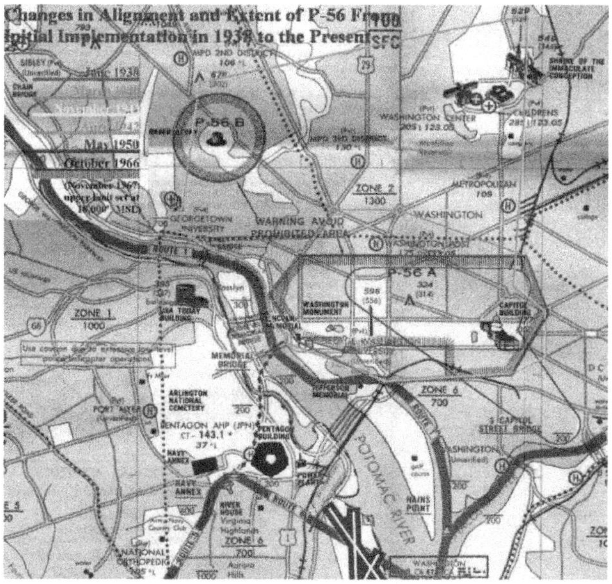

Answer to Question #3: The answer is choice A . . . maybe. (Actually it's A, definitely. My source was Bill Schumann, FAA Special Ops and Public Affairs, Washington DC.)

Answer to Question #4: It's choice A.

Answer to Question #5: The answer is choice D. I should know. I grew up on Long Island, under one of (then) Idlewild's approach paths.

Answer to Question #6: The answer is choice D. The next oldest pilot who ever lived incidentally, was Ralph Lang Charles. He worked for the Wright Brothers, knew Lindbergh, and flew an Aeronca Defender out of his own private airport located on his 25 acre property in Somerset, Ohio. Born in November 1899, he was flying until nearly his 102nd birthday. He died on February 2nd, 2003, at age 103. Obviously these things will tend to, ahem, change over time, but you can always ask the Information Systems Manager with the FAA, out in Oklahoma City, if you're ever curious as to who *really* has Seniority Number One.

Answer to Question #7: Choice D is nonsense, and there aren't any hard-surface runways that far up in northern Canada. The right answer is choice B. Choice C is wrong since by the time it's "off" it gets fixed.

Answer to Question #8: The answer is choice C.

Answer to Question #9: It's choice C folks: Victor 16. Someday, I'd like to file THAT one!

Answer to Question #10: The correct answer is choice B.

Answer to Question #11: The answer is choice C. Yup, just pull out your sectional. Several years ago, VFR charts started revealing this choice morsel of information at the bottom of the information area for each airport, for any pattern that is NOT left-handed. For example, for Frederick, MD, it now says "RP 5,12" (meaning of course, a right-hand pattern for runways 5 and 12).

Answer to Question #12: choice D. (If you consider structures only, then it's C.) The tethered aerostat radar system is used for drug interdiction, and has (or has had) active sites at Yuma & Ft Huachuca, AZ, Deming NM, Marfa, Eagle Pass, Matagorda, & Rio Grande TX, Morgan City LA, Cudjoe Key FL, and Lajas, Puerto Rico.

Answer to Question #13: choice D. Yes, really (at least, as of the date I wrote this). The oldest helicopter pilot was Mr. Johnny M. Miller of Poughkeepsie, NY. Dates of birth are not releasable, but Miller celebrated his 98th birthday at the

Centennial of Flight celebration in North Carolina. He flew until age 102 in fact. His "seniority number" was even low for fixed-wing pilots as well (of which he is also a member). In fact, he flew himself (solo) to Oshkosh in his Beech Bonanza, N19WC, in 2003. Not only has he been around for awhile, having spent over 80 years in the pilot's seat, but his rotary-wing flying also has an impressive span: he was flying autogyros back in the era of Amelia Earhart. He made the first transcontinental autogyro flight, in fact, in May of 1931 (beating her by about nine days, incidentally). He retired from Eastern Airlines in 1963, and often flew his Bonanza around the country to visit nine grandchildren and seven (at last count) great-grandchildren.

Answer to Question #14: The correct one is choice C. Lothar, younger brother of Manfred von Richthofen—Lord, yes, there was another one—had been a cavalry officer but transferred to the German Air Service in 1915. Considered by his older brother to be somewhat of a hothead, Lothar scored 24 victories in 47 days, and survived the war with 40 kills (only to die in an airplane crash in 1922). His older brother incidentally shot down exactly 80, twice that number of allied aircraft. (Also, Oswald Boelcke, choice D, who was the elder Richthofen's mentor, had the exact same number of kills as the Red Baron's younger brother.)

Answer to Question #15: Ladies first! It's choice B. On June 11, 1971, Shiela Scott of Great Britain became the first person (and one having the distinction of being female) to fly alone to the geographic North Pole in a small aircraft. (It was a Piper Aztec.) She published her autobiography "Barefoot In the Sky", in 1973. Not only did she fly over the pole, but she did so starting and ending at the equator. Admiral Byrd flew with Floyd Bennet in 1926 (and flew over the South Pole, though again not alone) three years later, though there is still some contention that Byrd and Bennet might have missed the actual North Pole by some distance (however he hit a bulls-eye with his Antarctic flight). Gus McLeod was the first person to fly to both the magnetic North Pole as well as the geographic North Pole (on April 17, 2000) however his distinction is that he was the first to do so in an open cockpit 1939 Stearman biplane! Roald Amundsen attempted to overfly the North Pole (and succeeded) in a dirigible in May 1926 (at about the same time as Byrd), flying from Norway to Alaska along with Lincoln Ellsworth and Umberto Nobile.

Answer to Question #16: It's C. In June 1986, American James Will hung out for 34 hours and three minutes over Oahu's Makapuu Point, a ridge enjoying nearly ideal "lift" conditions because of the trade winds. (The FAI however no longer considers hang glider endurance records for any subclass of hang glider.) And if you guessed choice D, come on now!

Answer to Question #17: choice D. Aye, folks do love a good story, and it's often easier to ask forgiveness than permission. Douglas Groce Corrigan was possibly the last of the early glory-seeking aviators to win the hearts and minds of millions before the world plunged into war. By then, Lindbergh's flight was eleven years old, and many had already flown the Atlantic, but it was still not something to try in a factory built airplane.

Corrigan's flying career began in 1925 in Los Angeles, when, for two dollars and fifty cents, he purchased a ten minute sightseeing ride in a Curtiss Jenny owned by B.F. Mahoney and Claude Ryan. Not long after, Charles Lindbergh arrived in San Diego with the dream that Ryan could build an airplane that would fly the Atlantic. Corrigan soloed in March, 1926, and he actually worked as a mechanic for Ryan in San Diego. He obtained his pilot's license in December 1927, and an air transport license three years later. In his autobiography, "That's My Story", Corrigan wrote that Ryan's crew were given 60 days to design and build the airplane, and were in such a rush that they often worked past midnight. (Yes, Corrigan actually worked on Lindbergh's airplane!) He made wing ribs, assembled the wing, installed gas tanks and lines, sewed, doped and installed fabric and helped install the instrument board. He barnstormed for a few years, and, in 1933, for about $325, he bought a single engine 1929 OX5 Robin. (He did later modify it for long-distance flight.) He worked sporadically in airplane factories in Los Angeles and (for Ryan) in San Diego. Corrigan actually spent three years trying to get permission to fly from New York to Dublin. He had been told that he could fly nonstop from New York to California, but that an ocean crossing was out of the question because his Robin was rather rickety and had no business challenging the Atlantic. (Also, Amelia Earhart had disappeared over the Pacific earlier that year.) Corrigan flew non-stop from California to New York in 1938.

At dawn on the foggy morning of July 17, 1938, he filed a plan for a non-stop return to California and departed New York's Floyd Bennett Field carrying two chocolate bars, two boxes of fig bars, a quart of water and a U.S. map with the route from New York to California marked out. A few puzzled onlookers watched him do a 180-degree turn and vanish into a cloudbank. About 28 hours later, he stepped out of his plane at Baldonnel Field, near Dublin, saying, "I just got in from New York. Where am I?" Whether or not he claimed that his compasses had failed, or he told officials that he had misread his compass and had meant to fly to Los Angeles, the Irish are much taken with a good yarn (especially when told by a transplant from the sod, and he was given a hero's welcome. He met the country's Prime Minister, the U.S. Ambassador to England, as well as Joseph P. Kennedy. Although Corrigan never quite admitted it, his error was surely a ruse to circumvent aviation authorities who had turned down his request to make the trans-Atlantic flight. No one believed Corrigan's explanation, especially the aviation authorities in both Ireland and America, who suspended his pilot's license and ordered his aircraft dismantled. Corrigan stuck to his story: He'd got turned around, his compass had malfunctioned, and he'd meant to fly to California. One of the statements that he is said to have made was indeed that "When I came down through the clouds I noticed I had been reading the compass needle backward." The only people who were fooled by that probably wanted to be.

He was the perfect anti-hero, and Corrigan's stunt caught the public fancy. By the time that the ship carrying Corrigan and his crated plane got back to New York, the suspension had been lifted, and "Wrong-Way" Corrigan was greeted as a hero. More than a million people lined New York's Broadway for a ticker-tape parade honoring the man who had flown in the face of authority. (The city gave him a bigger parade than it had given Lindbergh.) Corrigan enjoyed celebrity status for awhile and even played himself in the 1938 movie The Flying Irishman. So, yes, he really did know where the Hell he was going. He was no dope. (Who says Ronald Reagan invented plausible deniability?) Corrigan died in New York on December 9, 1995, at the age of 88. To most people, his name is still synonymous with football players who score goals for the opposition, but in reality, Douglas Corrigan was a beacon of passions and principles that shone above all obstacles to the realization of dreams.

Douglas Corrigan (l), and standing in front of the Curtiss OX5 Robin (r) in which he made his unauthorized 1938 flight to Ireland.

Answer to Question #18: choice B. On May 21, 1910 at Huffman Prairie, Wilbur and Orville flew together for the first and last time, with Orville piloting. (Also, on that same day Wilbur made his last flight as a pilot in the United States. He died in 1912.)

Answer to Question #19: Answer: Runway 13R-31L at JFK is listed as 14,572 x 150, and runway 14R-32L at ORD in Chicago is 13,000 x 200. Denver International Airport's new runway 16R/34L is 16,000 feet long. (It's the longest commercial runway in North America.) The Shuttle landing facility at the Kennedy Space Center is longer, being about three miles in length and quite wide, at 300 feet. The main runway 04-22 at Edwards Air Force Base in southern California, an occasional landing site for the space shuttle, is also one of the world's longest runways, at about 15,000 feet (see image). As for choice C, the mysterious 'Area 51' in Nevada, counting the 'over run' areas, it might be considered one of the longest at about 6 miles, but not all of that is believed to be paved. (Whether that's for the Aurora, the X-33, or what, is unknown.) Also at Edwards however are the Rosamond and Rogers Dry Lake beds, and its runway

17-35, the main Rogers Lake runway, is even longer, at about 7 miles.) The answer is choice A.

Note: I haven't counted 'landing lanes' at some sea plane bases; a few are several miles long. The Peekskill SPB in New York lists its 'runway' at 28,000 feet.

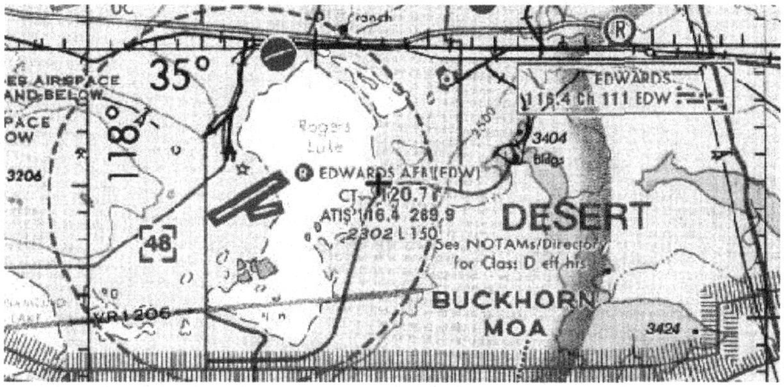

image of FAA Los Angeles Sectional Aeronautical Chart showing Edwards Air Force Base

Answer to Question #20: The answer is choice C; Tweety Bird, and it's the Portsmouth, NH (KPSM) RNAV (GPS) RWY 16 approach. ITAWT and ITAWA are two initial approach fixes; PUDYE is the final approach fix; and TTATT is the MAP. And of course, IDEED is the missed holding fix. (So that's: ITAWT ITAWA PUDYE TTATT, IDEED! There is also a HAMMM, BURGR and FRYYS in Vermont.) If you guessed B, shame, shame. Woody worked for a rival firm, but not WB.

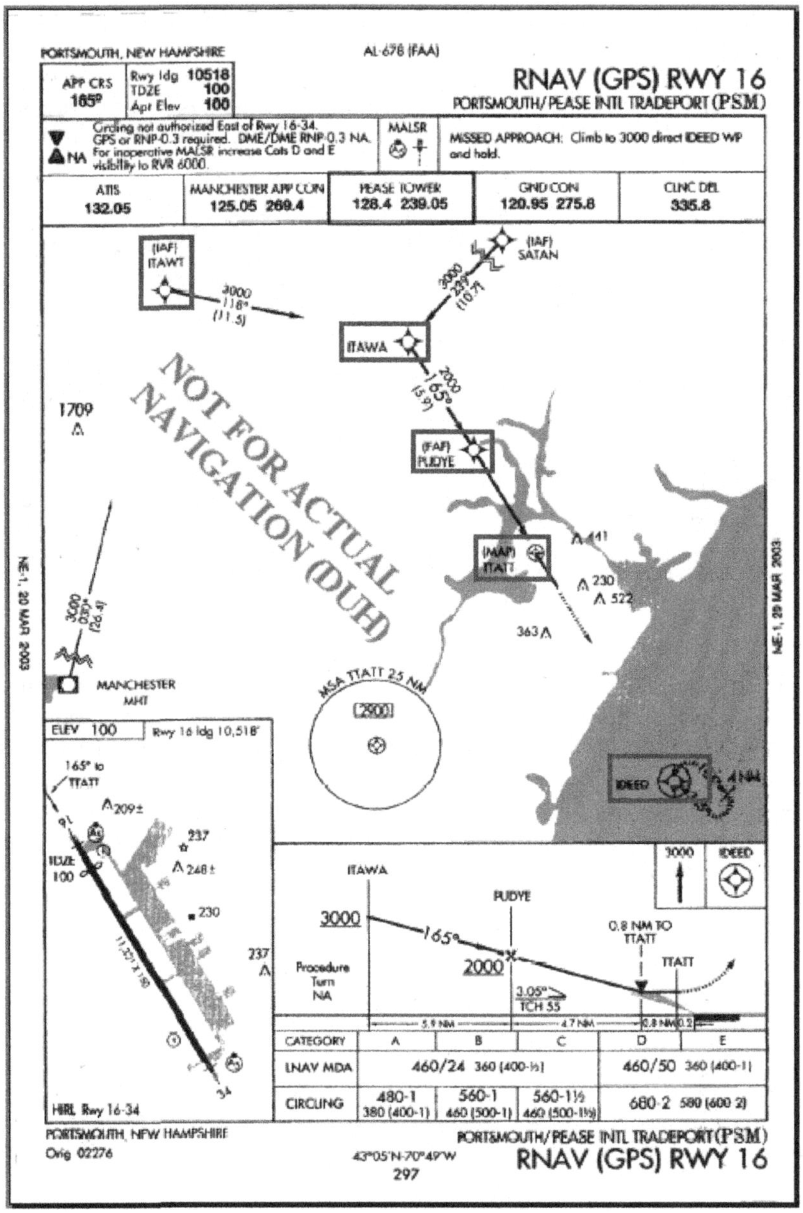

adaptation of FAA instrument approach procedure from U.S. Terminal Procedures Publication NE-1, RNAV (GPS) RWY 16 for Portsmouth, NH

Answer to Question #21: either choice B and D—strictly speaking, really just D. Although most sources refer to the faster edition of Yeager's "Glamorous Glennis" as the Bell X-1, it was actually called the XS-1, at the time. (This is how it was referred to in his actual flight transcript.) At 10:29 AM on October 14, 1947, over the parched and desolate lake beds of California's western Mojave Desert, Yeager, then a captain, flew the Bell XS-1. And exactly 50 years later, he re-enacted his milestone flight in an F-15 (at the time, his last flight piloting an Air Force plane) with Bob Hoover flying an F-16 chase plane. (Hoover had flown chase for Yeager during the original flight.)

Answer to Question #22: choice C. The year was 1910. The recipient of license #1 was Glenn Curtiss. The others (just four more) were: Frank Lahm (number two); Louis Paulhan (#3); followed by Orville and Wilbur Wright (numbers four and five, respectively). Do you see a pattern here? Glenn Curtiss was number one, and the Wrights were numbers four and five because, with so few recipients when the system of licensing began, they were assigned alphabetically!

Answer to Question #23: choice C. When this was written in late 2003, there were 7,484 public use runways in the US (14,968 if you count both directions), and 232 public-use helipads. 213 of those 7400+ runways are on bodies of water (at sea plane bases), and 23 runways are listed as "all ways" with no particular heading (including one rectangular landing surface listed as "balloon"). If you're interested in just what the distribution of public-use US runway lengths looks like, it looks like this:

As far as their lengths go...only 27 (less than 0.5%) are 1000 feet or less
467 (about six and a third percent) are greater than 1000 feet, and up to 2000 feet
1788 (about 24%) are greater than 2000 feet, up to and including 3000 feet
1879 (just over one-fourth) are greater than 3000 feet, up to 4000 feet
1338 (about 18%) are greater than 4000 feet, up to and including 5000 feet
874 (about 12%) are greater than 5000 feet, up to and including 6000 feet
403 (about 5.5%) are greater than 6000 feet, up to and including 7000 feet
232 (just over three%) are greater than 7000 feet, up to and including 8000 feet
152 (2.0%) are greater than 8000 feet, up to and including 9000 feet
150 (2.0%) are greater than 9000 feet, up to and including 10000 feet
61 (about 1%) are greater than 10000 feet, up to and including 11000 feet
49 (less than one percent) are greater than 11000 feet, up to 12000 feet
21 (less than one-half percent) are greater than 12000 feet, up to 13000 feet
11 (0.1%) are greater than 13000 feet, up to and including 14000 feet
24 (less than 0.5%) are greater than 14000 feet, up to 15000 feet
...and just eight (0.1%) are over 15,000 feet long

Significant numbers of the very longest runways are dry lake beds. One might easily notice from these numbers incidentally that very close to half of the nation's public-use runways are between two and four thousand feet long.

Answer to Question #24: choice C. The airplane was flown for the last time on April 30, 1928, after having accumulated a total of only 489 hours of flight. It was first put on display in the Arts and Industries Museum on May 13, 1928. It took its place in the then-new Air and Space Museum's "Milestones of Flight" gallery in 1976.

Answer to Question #25: choice D. Out of about 500 towered airports in the US, 75 have just one runway. (Also, 15 of these 75 have a runway longer than 8069 feet.) However, only three of these are 24-hour "big airport" operations. These are Lindbergh in San Diego, Pease in New Hampshire, and Harrisburg International. (There are two other "24/7" towered one-runway airports: one 2200-ft strip in Alaska, and one in the Mariana Islands near Micronesia.)

Answer to Question #26: It's D. According to the FAA database, it is owned by a student pilot in Houston, Texas.

Answer to Question #27: The right answer is choice B. I know: never heard of the guy, right? The first of these choices, the French WWI pilot Roland Garros, was actually the very first fighter pilot to be called an "ace". He was the first man to shoot down an enemy using a gun which fired through the arc of the propeller. (When he was shot down, the Germans captured his aircraft and discovered the secret of the plane that spat bullets through its propeller.) Garros escaped in 1918, only to be shot down later that year. At the time, France, Belgium, Italy, Russia, and the US used the term "ace" to mean five or more kills (although it is likely that Garros had only three). The German term 'kanone' signified a similar honor, but their minimum was 10.

The Red Baron was credited with 80 kills, and he was the highest scoring ace of WWI. Richard Bong was the highest-scoring US ace of WWII (or any other war, for that matter) with 40 kills. (Coincidentally, 40 was also the number credited to both the Red Baron's daring younger brother Lothar von Richthofen, as well as the more infamous elder brother's teacher, Oswald Boelcke.) However regarding lethality, the Germans seem to have had a corner on it in WWII. The German ace Werner Molders was the first to reach 100 kills (101, before he himself died in a crash). Many German pilots later reached 100, 15 reached 200 or more, and 2 scored 300 or more. Of these two, only one pilot in history reached the lofty total of 352: Erich Hartmann. (The other was Gerhard Barkhorn, with 301.) Hartmann, a German Luftwaffe Bf 109 pilot, reportedly shot down most of his opponents over the Russian Front, as well as other locations, such as Rumania. In about two and a half years beginning in October of 1942, he flew 1,425 combat missions and was shot down sixteen times, but he was never once wounded. When the war was over, Hartmann turned himself in to the Allies. They, in turn, turned him over to the Soviets, who tried him as a war criminal, and he spent over 10 years in the Gulag. Afterwards, he returned to West Germany. He died in 1995. He was devilishly handsome, and devilishly deadly.

Answer to Question #28: The right choice here is D. Yes, that far back. Central Airlines hired her, and she flew her first trip in 1934 (though just barely, on December 31). Her route was Washington, DC to Detroit, in a Ford Tri-Motor. However, she was the unwilling recipient of the most overt and blatant discrimination, was not actually given permanent status as a regularly scheduled pilot, and quit less than a year later. Jackie Cochran (B) was the first woman to break the sound barrier. Although there are reports of a female Aeroflot pilot in the 1950's, Turi Wideroe (A) was reportedly the first female pilot to fly an airliner on a regular schedule (i.e., not on some provisional or probationary status). She flew for Scandinavian Airlines. Ethyl Mertz (C) was a regular in the popular 1950's sitcom "I Love Lucy". (Actually, there was another early female airline pilot, Prem Mathur, who joined the Indian airline Deccan Airways in Hyderabad India, also in 1951. She flew DC-3s on regularly scheduled flights.) By the way, if this question looked familiar, give yourself a pat on the back, because a History Lessons question (Chapter 1) asked a question regarding which airline was the first to hire a female pilot.

Answer to Question #29: It's choice C. Harriet Quimby, born in 1875, actually started out as a newspaper writer. She was the first woman to fly across the English Channel, the first woman to fly at night, and (with just the slightest nod toward her gender) she also designed her own flying clothes. During the Harvard-Boston air meet in the summer of 1912, she was flying with a passenger in a Blériot airplane when her passenger made a sudden movement, dipping the wing of the aircraft, and they both fell out of the airplane from 2,000 feet above Boston harbor. (This was before seat belts.) Bessie Coleman was the first black female pilot (and also coincidentally died when she too fell out of an abruptly overturned airplane, in 1926). Katherine Stinson was the sibling of the famous Eddie of Stinson Reliant fame, and the fourth American woman to obtain a pilot's license. Her Ladyship the Baroness (choice A) was actually the first licensed woman pilot, period (on March 8, 1910). Harriet followed her, though, not long after (August 1, 1911).

Answer to Question #30: It is choice C. On June 17 and 18, in 1928, Amelia Earhart became the first woman to cross the Atlantic. (However, she was only a passenger!) To her distress, the media gave her all the attention. She *did* cross the Atlantic alone on May 20-21, 1932 in a Lockheed Vega, becoming the first woman to fly solo across the Atlantic from west to east (from Newfoundland to Great Britain). In 1936, Beryl Markham became the first woman to fly across the Atlantic from east to west, flying from England to Cape Breton Island, Canada. Jackie Cochran was the first woman to fly faster than the speed of sound (with a little help from her friend Chuck Yeager) in an F-86 Sabre, in 1953.

Answer to Question #31: Both choices A *and* C. In general, only US Government aircraft can go supersonic over land. (But there *are* provisions for civil aircraft to exceed Mach 1 in Appendix B to CFR Title 14, Part 91.817, and such instances usually involve a darned good reason for some entity in the private sector to do so.) As to where NASA and Uncle Sam's best boys and girls can give their airplanes a workout, one example is the 17-mile wide and 240-mile long Black Mountain High Altitude Supersonic Corridor, passing over Edwards AFB in California and extending towards Nellis AFB near Las Vegas, Nevada, over the Lake Mead National Recreation Area. It extends from Ventura County, California to Clark County in Nevada, and passes through portions of several other counties. It overlaps the R-2508 complex near EAFB, across the northern section of R-2515. (Actually, there are three designated corridors in that general area.) USAF aircraft going supersonic usually notify either the FAA High Desert Terminal Radar Approach Control facility (call sign JOSHUA) or the Air Force Flight Test Center Radar Control Facility (call sign SPORT). Also, these aren't just Top Gun jocks doing it because they feel like a Need For Speed. Each training base has a Letter Of Agreement on file with the FAA, and maintains a "boom log" to keep track of every supersonic flight. Additionally, authorized aircraft *are* actually permitted to go supersonic anywhere in the restricted airspace at Edwards, if they need to. They must however remain sensitive to the impact that may have on people who live beneath their route of flight. In the wake of the terrorist attacks on September 11 2001, F-16s on combat air patrols *have* exceeded Mach 1 (choice A), in order to catch up to airliners during air-rage incidents. But during most non-emergency peacetime conditions, overland supersonic flight activity is usually above 30,000 feet. (See USAF Instruction 11-202, Volume 3, Chapter 5, Section 5.7.1, and AFI 13-201, Supplement 1, Section 2.11, "Supersonic Operations".)

As to choice D, whether a sonic boom is heard or not is a function of air pressure and temperature, humidity, load factor on the aircraft, terrain, Mach number of the aircraft, planform, inlet type, etc. According to Marta Bohn-Meyer (to whom I directed this specific question regarding any limiting altitude for sonic boom generation), even from 80,000 feet, sometimes the sonic boom from the SR-71was hardly noticed on the ground...and sometimes, it was incredibly strong. (Marta has, as she puts it, the distinction of being the world's only SR-71 crewmember who happens to be female. She has also conducted research on exactly this subject for NASA.) Here is what the vicinity of the Black Mountain corridor would look like on a sectional chart, followed by its actual depicted location:

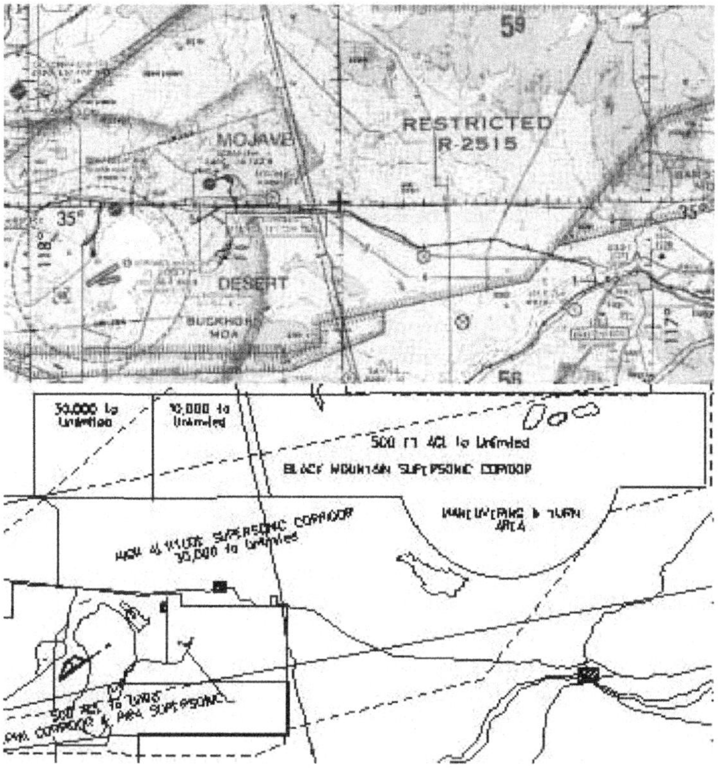

Top image: the Black Mountain High Altitude Supersonic Corridor from Edwards AFB in California towards Nellis AFB near Las Vegas, Nevada, from an FAA Sectional Aeronautical Chart. Bottom image: from the USAF Instruction 11-202, Volume 3, Chapter 5, Section 5.7.1, and AFI 13-201, Supplement 1, Section 2.11, titled "Supersonic Operations".

Answer to Question #32: B. N60094 is a Bleriot XI monoplane, manufactured in 1909 (and listed in the Experimental category). It belongs to the Rhinebeck Aerodrome Museum. (There are two other Bleriot airplanes in the United States; one is in California and the other is in Oklahoma, but these are not flying.) It is the second oldest airplane in the world that is still airworthy. (However, according to information obtained from the Aerodrome, only the front and rear third are original; the wings, elevators, and horizontal stabilizer have been rebuilt.) The original aircraft had crashed in 1910; its remains were later removed from a junkyard and were stored for many years before Cole Palen returned it to flying condition. It is powered, as was the original, by an Anzani engine.

Answer to Question #33: choice B. The southernmost *paved* runway in the Americas, as well as the entire world, is Guardiamarina Zanartu, or in ICAO-speak, SCGZ. It's 4700 feet long, and located in the town of Puerto Williams, the capital of the Chilean Antarctic Territory, on the south side of the Beagle Channel, in Navarino Island (Isla Navarino). It is also considered the southernmost port of the world, with a naval base and a main commercial center in the area. Although it has a small population of only two or three thousand people, it has a hostel, an airport, communication systems, a museum, and a yacht Club. Just to the north on the other side of the Beagle Channel lies Ushuaia (SAWO), the capital of the Province of Tierra del Fuego, in the Argentine Republic (more than a bit further south than Christchurch, New Zealand). The latitude is about 55 degrees South (versus that of Christchurch, which is about 43.5 degrees S or Capetown, South Africa at about 34 S). Although Guardiamarina Zanartu has Ushuaia beat for the title of "southernmost paved airport" by just a few miles, Ushuaia is significantly larger and is the usual jumping-off point for flights to Antarctica. The Punta Arenas airport is the world's most *southerly commercial airport (at* about 53 degrees South). There *is* a runway in Antarctica--actually, there are about 40+ landing facilities at different locations operated by over a dozen governments, with runways that are either gravel, sea ice, glacier ice, or compacted snow (and no paved runways).

Antarctica is the world's highest and driest continent (and technically, the world's largest desert). Near the biggest manned station, McMurdo, which is at about 78 degrees S, there are three airfields and one helipad. Originally there was one ICAO identifier for McMurdo Station: NZCM, which stood for Williams Field at McMurdo. (Because all intercontinental flights come from New Zealand, they were assigned an "NZ" ID.) They recently had new identifiers registered with NOAA for reporting weather observations. NZIR is for the "Ice Runway" which is close to McMurdo, is built on annual sea ice, and is used at the start of the season; they can land anything from Twin Otters to C-5's on it. After the ice starts to melt, operations are moved to Williams Field (NZWD), on the Ross Ice Shelf about eight miles from McMurdo; operations there are limited to aircraft equipped with skis (Twin Otters and LC-130's). The third runway is the Pegasus runway (NZPG), another blue ice runway which is used at the end of the season, and which is located about 12 miles from McMurdo and supports the "heavies", or wheeled aircraft. The helipad is located right at McMurdo Station. (The runways on ice at McMurdo are the only ones for wheeled aircraft that are operated by the U.S in Antarctica.) The very most southerly airport in the world is at the Jack F. Paulus Skiway at the Amundsen-Scott Base, at 89° 59' 59.99" South (and an elevation of 9300 feet) with an identifier of NZSP. You can sometimes find data from these stations on the internet. (In case you are wondering, that one one-hundredth of one second of latitude away from exactly 90°S amounts to all of about one foot. This is puzzling since the wandering of the poles due to nutation alone, also known as the "Chandler wobble", exceeds that by at least an order of magnitude...)

Below is a picture of the Ice Runway at McMurdo, kindly sent in by my most southerly correspondent, Bob Vehorn, who has been living in Antarctica for one month during each of the last several years. McMurdo is the current headquarters of the United States Antarctic program, and is home to well over 1,000 people during the Antarctic summer. (This comprises about a fourth of the continent's entire summer population.) During their winter, that drops to about 200 people. (That's Bob in the picture, with NZIR behind him, and on the far right is the hut that Captain Scott built during the Discovery expedition of 1902.)

personal photograph of Bob Vehorn, taken at McMurdo Station, in Antarctica, reproduced with permission

Answer to Question #34: It's C. If you knew that it's based on what fits in those little bitty circular runway symbols, you were on to something there. Choice B is close. (Also, choices A and D are total fiction.) As you know, different aerodrome symbols are used to depict runways on various aeronautical charts, such as Terminal Area Charts, and Sectionals. Several different symbols are used to show hard-surfaced runways. First, regarding sectional charts, here are the gory details, according to the National Aeronautical Charting Office, or NACO (whose 300 or so employees were transferred from NOAA to the FAA on October 1, 2000). When a runway is between 7970 and 8069 feet in length, they say that cartographers round it to 8000. This equates to a line that is 0.192 inches in length on the sectional's scale. In this case, a circular aerodrome symbol is used. (They're pretty good about showing the relative orientation of the predominant runway, and some effort is made to proportionally shorten the line a bit for runways that are between 1500 and 8069 feet long.) NACO says that if the runway is between 8070 and 8169 feet in length (inclusive, incidentally), they round to 8100, which equates to a line 0.1944 inches in length on the sectional's scale. Since this line is too long to fit into the largest circular aerodrome symbol NACO has available (which happens to be 0.193656 of an inch, which when multiplied by the sectional's scale factor of half a million, comes out to 8069 feet), they place a box around the actual runway pattern forming a representative enclosed shape. So when a runway is 8070 feet or longer, it's going to actually look like a smaller version of the runway, right? Well, almost. Actually, for airports with multiple runways, if the maximum distance between the two furthest points between the respective opposite ends of any two runways exceeds 8069 feet (even if the longest runway length is shy of 8069), then they do it that way, too. And for Terminal Area Charts, outlined boxes showing the actual runways are used for any airport with a runway that is just 1500 feet long, or longer. (Why isn't the TAC chart cutoff 4035 feet, since it shows things at twice the scale of a sectional? Probably because for local flights where pilotage and perspective matter more, they want us to have a better idea of what to look for.) This figure below is fairly representative of what you will see on a sectional:

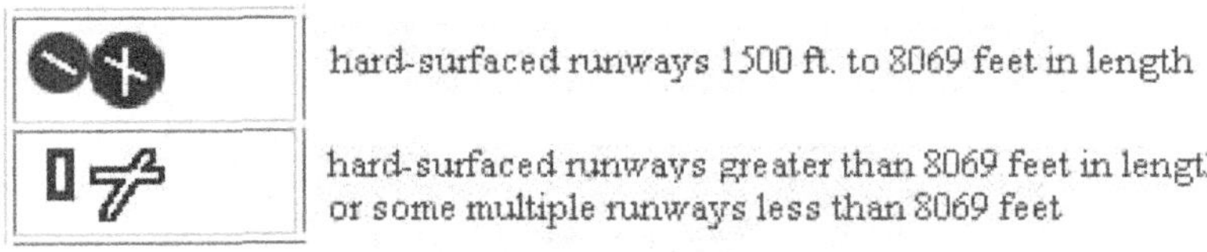

Answer to Question #35: choice B. The fly-in community of Mountain Air, located on the top of Slickrock Mountain outside of Burnsville, North Carolina has a 2900 x 50 foot runway that is 4,432 feet above mean sea level, and is not only the highest airport east of the Mississippi but (*they* say) east of the Rockies, as well. This claim could be a stretch, since there are several airports in western Nebraska, Texas, and South Dakota which are close to, or over, a mile high, such as Custer County in South Dakota, at 5,602 feet MSL. (Kimball Municipal in Nebraska is at 4,926 feet, and in Texas, Marfa Municipal is at 4,849 feet MSL. It really depends on your definition of what "east of the Rockies" really means.) Anyway, Mountain Air's country club is at 4390 feet, and they have a spectacular "airway fairway" (golf course). The pattern altitude is 5400 feet MSL. North Carolina's highest *public-use* airport, according to the AOPA database, is Ashe County in Jefferson NC, at 3180 MSL, and according to the same database, the highest public-use airport in the East is Ingalls Field in Hot Springs VA, at 3792 feet. Of course, it depends on what you mean by "the east"...

As far as highest runways in the *Far* East, or anywhere else on the planet, Changdu Bangda airport (ZUBD) in eastern Tibet (not Georgia) is currently the world's highest public-use airport, at 14,219 feet above sea level. (Its runways are over 18,000 feet long.) La Paz, Bolivia's JFK (SLLP), used to be called the world's highest airport at 13,313 feet. Lhasa's Gonggar airport in Tibet's capital (ZULS) is still a respectable 11,621 feet. Regarding choice C, North Carolina's *Mt.* Mitchell, at 6,684 feet, is the highest point east of the Mississippi, and as far as choice D goes, there *is* no Albuquerque in Tennessee. (However, there *is* a Mountainair Municipal that is actually in Albuquerque New Mexico, at 6492 feet, as well as Angel Fire airport, at 8380 feet.)

Answer to Question #36: It's choice C. The XF-84H was modifed from an F-84F "Thunderstreak" jet, in a retrograde sort of way, by instead using a turboprop engine (which was located behind the cockpit, incidentally) and converting the production tail to a T-tail. Two airframes (S/N 51-17059 and 51-17060) were modified to accept Allison XT-40-A-1 engines. These aircraft, designated XF-84H, had special propellers *designed* to rotate at supersonic speeds. Between 22 July 1955 and 9 October 1956, the two XF-84H prototypes made twelve test flights. Eleven of the twelve flights ended in emergency landings. Sounds produced by the aircraft's turboprop engine--actually, it was the propeller--caused nausea and headaches among ground crews, earning the XF-84H the unofficial nicknames "Thunderscreech" or "Thundershriek" because of the intolerable sound made by the sonic booms from the propeller, which made it basically unusable. The propeller blades went supersonic *even while the airplane was running up on the ground.* This was definitely the noisiest *single-engine* aircraft ever to fly. It wasn't just the volume, and it wasn't *sound* as we think of it, but shock waves, which are typified by a nearly instantaneous pressure increase. A person standing some distance from the airplane was subjected to rapid fire shock waves. These shock waves acted directly on the large intestine, causing spasms that caused individuals in too-close proximity to soil their pants. Hearing protection made no difference.

The Hornet (choice A) was the first production tip-jet powered helicopter. There was basically no engine inside the fuselage other than a small one-horsepower engine to get the rotor up to speed. This was so that the two Hiller 8RJ2B Ramjets, which were mounted at the tips of the rotor blades, could begin to function. Although the two 12.7-pound ramjets only produced 38 pounds of thrust, they were horrifically noisy. And though the aircraft was tiny (empty weight 544 lb) and looked like something out of a 1960s television cartoon (see photo), it was probably the noisiest helicopter ever made. The Ba-349 Natter (choice B) was a disposable, vertically launched rocket-powered attack plane used by a desperate Germany early in 1945, and the Buffalo (choice D) was a pudgy and ineptly designed fighter at the start of WW II.

Republic XF-84H Thundershriek, on the ground, and in flight
(courtesy of the US Air Force Museum at Wright-Patterson AFB)

Hiller YH-32 Hornet, at rest and in flight (or is that George Jetson?)
bottom photograph courtesy of the Museum of Flight in Seattle, Washington

Answer to Question #37: choice D (pretty anticlimactic, wouldn't you say?) First, Class F airspace, in those countries that use it, can differ from the ICAO standard. Amendment No. 33 (there are about 41) to Annex 11 (one of about 18 to the 1944 ICAO Convention, in this case, on Air Traffic Services) explains ICAO airspace classifications. The Preamble to CFR Title 14 FAR Part 71 summarizes them. Anyway...ICAO Class F airspace is uncontrolled airspace in which a separation service (often called an ATS or air traffic advisory service) is provided to *participating* IFR flights. If IFR flights choose to call the ATS, they will be separated from other IFR flights who are also in contact. There is no requirement to obtain an ATC clearance in class F airspace under IFR, *or* VFR. The UK implements Class F, pretty much by the book, for standard routes in relatively inactive airspace. In other countries, Class F airspace is uncontrolled airspace in which an advisory and separation service is available for IFR traffic, and in which a Flight Information Service is provided between VFR and IFR as well as among

VFR traffic. Canada uses Class F. Some countries, such as New Zealand, do not use Class F except where they cannot, as they say, provide a normal ATC service for natural or human reasons. (No, I have no idea what that might mean, either.)

Answer to Question #38: It's choice B. Jacqueline Auriol was a decorated French pilot who in 1953 became the *second* woman to break the sound barrier. She started as a stunt pilot and went on to be a military test pilot, flying more than 100 types of planes. The exact opposite in terms of how she started out in life, she was the daughter of a wealthy shipbuilder and timber importer. In 1938 she married Paul Auriol, the son of Vincent Auriol, a prominent leader in the Socialist party and the former president of France. During World War II, Madame Auriol, by then a mother of two sons, evaded Gestapo agents and assisted the French Resistance. She took up flying in 1947, and started stunt flying. In 1950, she obtained a military license and qualified at the Flight Test Centre at Bretigny, France, as (if you believe the French references) the world's first woman test pilot. In May 1951, she set a new women's speed record in a British Vampire jet, flying over 508 mph and beating Jackie Cochran's previous record, set in a P-51. This began a friendly rivalry between the two ladies, and they traded women's speed records until 1964. Although she suffered severe injuries during an airplane crash in 1949, she resumed flying in the US and qualified to fly jets. Flying a Dassault Mystere IV on August 3rd, 1953, she became the second woman to break the sound barrier. (After 1959, their competition continued in excess of Mach 2.) Mme. Auriol was also one of the first pilots to fly the supersonic Concorde. Jacquelyn S. Parker (choice A) was the first American woman to become combat qualified in the F-16. Graduating from college at just 17 years of age, she was already a "fast burner". In 1988, then Capt. Parker became the first woman pilot to attend U.S. Air Force Test Pilot School at Edwards AFB. The name in choice C was intentionally fictitious, and Jacqueline Bouvier, later known as Jackie Kennedy, and then Jackie Kennedy Onassis, was never a pilot.

Answer to Question #39: The right answer is choice C. The Brewster Buffalo was the Navy's first monoplane carrier-based fighter, in 1939, and it had four one-half inch machine guns. (It also flew pretty much like a buffalo.) There never was a Grumman F-13. (Curiously, the Navy's F-14 *would* have received this number, but the number 13 was rejected by Grumman and/or the Navy, obviously for superstitious reasons. Since the late 1960s, it has been an unwritten rule that "-13" designators are always skipped.) The Fokker F-10 Trimotor was an airliner, back in 1927, and it enjoyed success up until the widely publicized crash that took the life of football coach Knute Rockne on March 31, 1931. The F-117A is designated as a fighter (subsonic, incidentally) but it has no guns, and usually, no missiles. The reasons are kind of obvious, when you think about it. And in case you were wondering about those designators...

Beginning in 1919, the US Army Air Service assigned 15 simple alphanumerical designations to all its aircraft. The basic system, consisting of letters representing its primary design task, followed by a sequential "design" number, was continued by the US Army Air Corps, US Army Air Force, and the newly created branch of the US Air Force, even though the representative letters have significantly changed, several times (such as in 1924, and between 1924 and 1948, when the Army and later the Air Force made further changes). One of these involved dropping the "A" (for attack) and "P" (for pursuit) and instead using "F", for fighter. (The letter "P" has been used for "patrol, and the "A" has been used as a mission modifier, however.) In 1962, this designation system became mandatory for use by all U.S. military services.

Answer to Question #40: I wondered about that for the longest time. I finally got my answer from Seth Shulman, the author of the Curtiss biography "Unlocking the Sky". As he told me, the answer was found by Curtiss' *first* biographer, Clara Studer. Glenn Curtiss' mother gave him the middle name Hammond after her beloved hometown. (In addition, she decided to name him Glenn after a visit to the dramatic Watkin's Glen, not far from her home.) So it's C.

Answer to Question #41: Answer: Yes, there really were two "Wild West" showmen with the same last name (and with the same basic image, down to the mustache, goatee, and ten-gallon hat. Actually, the truth is that Sam Cody was born with the surname Cowdery, but changed it in order to emulate (and let's be realistic, also possibly profit from) his hero, Buffalo Bill. Although Sam Cody did develop an interest in man-lifting kites (and later, dirigibles) for purposes of aerial observation during times of war, neither one did any commuting, per se. (Incidentally, he was also born in the town of Birdville, Texas.) Harry Houdini (born Erich Weisz) however, did use the airplane for the very utilitarian purpose of getting from one place to another.. In 1910 he bought a Voisin biplane and learned to fly it. He used it to fly himself from town to town and keep his entertainment commitments. He was also the first person to fly an airplane in Australia. The right answer is choice C.

Answer to Question #42: Before I blurt out the answer, first, an unsolicited word about some technical minutiae: the ducted fan. It offers advantages at both high and low subsonic speeds. Most of us have flown on airliners powered by turbofan engines. Although the lower speed realm for the most part comprises obscure VTOL and air-cushion vehicles, or perhaps the enclosed fenestron tail rotor of some turbine helicopters, it is the lower speeds where the advantages of a ducted fan over a propeller in the open free stream of air are actually the greatest. With a ducted fan, there is no region of reversed flow near the propeller tip, there is a higher net mass flow of air and significantly greater static and low-speed thrust, additional lift force using a properly shaped duct, less propeller tip "buzz" noise, and of course, the opportunity for thrust vectoring. Now we're all familiar with the appearance of a ducted fan turbine, and certainly having jet engines placed *inside the main fuselage* of jet fighter aircraft, reminiscent of the primitive digestive system of an annelid worm, wherein air is taken in through the front and then expelled straight out of its posterior end at high speed, which draws no second looks. But what about doing that with a reciprocating engine *and a propeller*?

In the late 1920's the Italian Air Ministry designed an experimental aircraft with a tubular "venturi duct fuselage." Designed by Luigi Stipa, in 1932, Gianni Caproni's company built one. It stood about 11 feet tall, had a length of about 20 feet, and a wingspan of almost 47 feet. It was powered by a 120 hp D.H. Gipsy III engine. The top speed was 83 mph, and the lift effect of the tube made the landing speed 42 mph. (Yes, the answer is B.) It might be said to resemble a kind of fascist Gee-Bee racer (although the fuselage width to overall length ratio of the Stipa was actually almost twice that of the Gee Bee). Although the height-to-length ratio for the "Goblin" (choice A) was about the same as the Stipa, it was jet powered. As for the GZ-20 (choice C), it does have quite an impressive girth, yes, but still its stoutness doesn't match these other two. It has a width of some 50 feet, but its length is over 190 feet. Of course, you would expect that of the Goodyear Blimp... As for the SG (or the newer SGT-F) in choice D, that's the Super Guppy. Although it's a zaftig 25 feet wide and the fuselage is 32 feet high, with a length of about 144 feet, it's hardly a competitor.

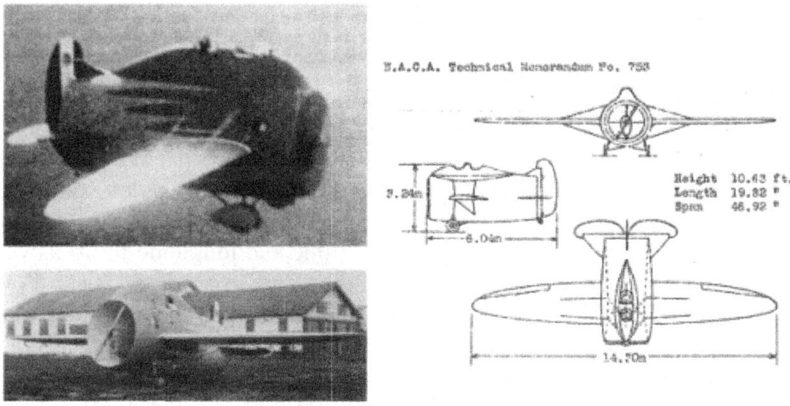

photographs and diagrams of the Italian Stipa monoplane, ca. 1932;
with permission from Sequoia Aircraft Corporation

Answer to Question #43: Some things in aviation open up a can of worms. This is one of them. Glider pilots used to do it all the time (those who had at least rudimentary instruments for attitude control, that is). In Gerry Casey's book, "Flying As It Was" he mentioned flying gliders up past the ceiling, and right up into clouds. However, that was half a century ago. Today, it's not easy to pin down just where, when, and how someone might do this in an un-powered glider. Just what regulations in Part 61 or Part 91 address testing, qualifications, or instrumentation requirements that would allow a pilot to continue climbing in a thermal, right up into a cloud?

First of all, there are specifications for instrument ratings in airplanes, helicopters, and powered-lift aircraft, but you won't see anything about gliders (or lighter-than-air vehicles either, at first). As it turns out, they're covered, but as is typical of the regulations, the coverage is oblique, and inferential, and the interpretations can be pretty nebulous. Let's start with CFR Title 14, Part 61.3(e)(3). According to this ruling, it looks like all one needs are glider and instrument ratings:

> "(3) For a glider, a pilot certificate with a glider category rating and an airplane instrument rating; or..."

So yes, there *is* no such-a thing as a glider instrument rating, folks (nor for airships). It would seem that all a pilot needs to have is a current instrument rating and be in an IFR-equipped sailplane. (Actually, the pilot would also have to be instrument-current in gliders, and there's actually a regulation addressing this.) Continuing my research, I called my local Flight Standards District Office, my local friendly Designated Examiner, my local Washington DC area Tracon ("Potomac"), AOPA's Technical Services folks, as well as the Denver ARTCC (out where wave soaring can go to phenomenal heights) and the Soaring Society of America out in Hobbs, New Mexico. I also spoke with noted sailplane authority and author Tom Knauff in Pennsylvania, as well as with Frank Phillips, ATP and an FAA Designated Examiner (as well as an attorney) in Easton, Maryland, and finally, noted aviation writer, retired airline pilot, and author, Barry Schiff. Here's what I was able to piece together.

First of all, back East, there is virtually no soaring in IMC. (Tom Knauff asserted in fact that in the US, there isn't any at all, period, but that it's often done in England.) Out where the sky is big and the orographic lifting potential is much greater, there are actually charted soaring areas, particularly for wave soaring, which are blocks of reserved airspace with latitude and longitude limits as well as floors and ceilings. For example, with the Pikes Peak (so called "Black Forest") areas 1, 2, 3, and 4, they go from FL 180 all the way up to FL 600. There are others as well, such as Salida (1 through 5), or Curley Peaks. According to a Denver Center controller, there are Letters of Agreement with several soaring societies, as well as the USAF Academy. (However, these are more for penetration into Class A and not into IMC. Also, IFR operations in Class A airspace can become hazardous for aircraft such as gliders, because of the dangers of icing.). A pilot must call several hours in advance to make such a reservation, and they ask pilots to remain in two-way communication. Apparently there's some degree of contrast with FAR 91.205, because transponder use is not mandatory (or at least, observed). The IFR rules for gliders are the same as for airplanes. As you might guess, most sailplanes lack many of the instruments that are required for IMC flight under this rule, and most do not even have a radio. (In the case of a glider so equipped though, all inspections for those instruments must have been performed with in the preceding 24 calendar months. There also used to be an Advisory Circular 20-46, effective back in September of 1965, titled "Suggested Equipment for Gliders Operating Under IFR", but it has since been removed.) In controlled airspace—this would mean Class E, most likely—you need a flight plan, and a clearance (and a block of airspace). One helpful aspect would be to find a slow sector, traffic-wise. Otherwise, you are limited to IFR flight in uncontrolled airspace. According to Barry Schiff, a glider pilot can fly IFR in any uncontrolled airspace that has been blocked out for that purpose. (I don't know about you, but swooping around inside a cloud in a non-radar environment isn't something I'd feel real comfortable doing.) But, is there anything at all in the "FARs" that sanctions or even refers to flying a glider in IMC? You bet! It's right here, in Part 61.57, in sub-part (c)(2), in fact:

§ 61.57 Recent flight experience: Pilot in command.

. . .

(c) Instrument experience. Except as provided in paragraph (e) of this section, no person may act as pilot in command under IFR or in weather conditions less than the minimums prescribed for VFR, unless within the preceding 6 calendar months, that person has:

> (1) For the purpose of obtaining instrument experience in an aircraft (other than a glider)...
>
> (2) For the purpose of obtaining instrument experience in a glider, performed and logged under actual or simulated instrument conditions --
>
> > (i) At least 3 hours of instrument time in flight, of which 1 1/2 hours may be acquired in an airplane or a glider if no passengers are to be carried; or
> >
> > (ii) 3 hours of instrument time in flight in a glider if a passenger is to be carried.

The bottom line? As FAR (hah, hah) as I can determine, it's B.

Answer to Question #44: ATC can issue advisory information on bird activity. And this example uses the exact phraseology recommended in the controller's handbook, or FAA Order 7110.65N. It's choice C. Here is the paragraph, in fact, verbatim:

2-1-22. BIRD ACTIVITY INFORMATION

> **a.** Issue advisory information on pilot-reported, tower-observed, or radar-observed and pilot-verified bird activity. Include position, species or size of birds, if known, course of flight, and altitude. Do this for at least 15 minutes after receipt of such information from pilots or from adjacent facilities unless visual observation or subsequent reports reveal the activity is no longer a factor.
>
> ***EXAMPLE-***
> "Flock of geese, one o'clock, seven miles, northbound, last reported at four thousand."
> "Flock of small birds, southbound along Mohawk River, last reported at three thousand."
> "Numerous flocks of ducks, vicinity Lake Winnebago, altitude unknown."

b. Relay bird activity information to adjacent facilities and to FSS's whenever it appears it will become a factor in their areas.

Answer to Question #45: These are only in the high-altitude en route structure for use by aircraft with the proper equipment. That includes those with the following suffixes: /E (FMS with en route, terminal, and approach capability); /F (FMS with en route, terminal, and approach capability that meets the equipment requirements of /E); /G (Global Positioning System/Global Navigation Satellite System or GNSS-equipped aircraft with en route and terminal capability); /R (Required Navigational Performance), etc. You might wonder: why bother with restricted routes for RNAV capable aircraft, in the first place? The FAA published over 150 such routes in the 1970's, but the FAA discovered (surprise, surprise) that most RNAV aircraft were using RNAV on a random route basis. They revoked them early in 1983 (except for four such routes in Alaska), but in April 2003 the FAA adopted ICAO terminology and in May, published 11 new RNAV routes in the US domestic high-altitude structure. The answer is choice C.

Answer to Question #46: It could be Saint Patty's Day, but that's not the excuse. If two grades were accidentally mixed, they'd come out clear, not green. And it's doubtful that automotive antifreeze would find its way into your fuel tanks. The answer is choice A.

Answer to Question #47: Most places, yeah, but not everywhere. Strictly speaking, the answer is: false. Check out CFR Title 14, Part 71, and you'll see why. For example, over Hawaii, would you like Flight Level 300? There is no Class A over Hawaii, so...knock yourself out! Here's that FAR:

> § 71.33 Class A airspace areas.
>
> (a) That airspace of the United States, including that airspace overlying the waters within 12 nautical miles of the coast of the 48 contiguous States, from 18,000 feet MSL to and including FL600 excluding the states of Alaska and Hawaii, Santa Barbara Island, Farallon Island, and the airspace south of latitude 25°04' 00" North.
>
> (b) That airspace of the State of Alaska, including that airspace overlying the waters within 12 nautical miles of the coast, from 18,000 feet MSL to and including FL600 but not including the airspace less than 1,500 feet above the surface of the earth and the Alaska Peninsula west of longitude 160°00' 00" West.
>
> (c) The airspace areas listed as offshore airspace areas in subpart A of FAA Order 7400.9K (incorporated by reference, see § 71.1) that are designated in international airspace within areas of domestic radio navigational signal or ATC radar coverage, and within which domestic ATC procedures are applied.

Answer to Question #48: You probably eliminated choice B (1974) immediately. (Surely someone must have complained about loud noises before then, right?) And choice A, 1949, so soon after the dawn of the jet age, might seem a likely choice. (You also most likely suspected choice C, 1907, as a red herring; that's just a tad *too* early!) When you think about it, the most plausibly realistic choices would be A or D. The fact is, it's choice D (1928). The owner of—and I'm not making this up—the Cackle Corner Poultry Farm in Garrettsville, Ohio registered a complaint to the Postmaster General, informing him that the noises of low-flying airplanes was unsettling his chickens and putting a big dent in their egg production. The complaint was forwarded to the offending party (the National Air Transport, a local airline) with the recommendation that they pay more attention to maintaining higher altitudes when flying over the farm.

Answer to Question #49: The Wright brothers may have astounded the nation early in the 20th century, but they also had to overcome more than just a little public prejudice. Even though mankind had dreamed of flight for ages, when most people came face to face with aviation in all its cacophonously clattering infant glory, some were just as likely to run the other way as they were to simply stand dumbfounded and gaping, open-mouthed. They knew that if they were going to win public acceptance, they would need more pilots to show just what airplanes could do. In addition to exhibitions, that also meant flight schools. (So Orville and Wilbur were, in a manner of speaking, the creators of the world's first "Be A Pilot" program!) The very first pilot that the Wright brothers hired was one Walter Richard Brookins. (I know what you might be thinking: "Who?!") He was a native of Dayton, Ohio, as well a long-time friend (as in, since the age of *four*), and had been a student of their sister Katharine's, at Central High. He attended the Wrights' flying school near Montgomery Alabama, soloed in only two and a half hours, and became their very first instructor. (True, somebody had to teach him, and since that somebody had been Orville, there might be *some* argument for choice A.) His job was to train pilots for the new Wright Exhibition Team. In the year that he became the first "civilian pilot" he set records for altitude, becoming the first pilot to fly an airplane a mile high (actually, 6175 feet) and was the first pilot to fly at night. He was also the first pilot to demonstrate aerobatics. Yes, it's D. By the way, his portrait is one of the thirty or so enshrined in the main hall of the Visitor Center at the Wright Brothers National Memorial, and you'll find this story right beneath his image, much as I've related it.

reproduced with permission from First Flight Shrine, Wright Brothers National Memorial and National Park Service - Outer Banks Group

Answer to Question #50: According to information from the Civil Aviation Registry's Airmen Certification Branch in Oklahoma City, on one particular recent day, there were exactly 627781 active US pilot certificates. And at that same point in time, there were exactly 47,332 controller certificates in the United States. However, most of them are not active and the number is believed to be more like 17,000. (So pilots outnumber controllers by, at a minimum, well over an order of magnitude.) But the statistic under discussion here is that (according to the same source) there were at that same point in time approximately 4,060 controllers with active pilot certificates. Although probably not all of those controllers were also *actively working* as controllers, *most* of them probably were, so that would put the proportion of working controllers who know both sides of the microphone at about one quarter (choice C).

Chapter 4: Airmanship and Aerodynamics

This is where you are likely to see questions which, in addition to the occasional astounding factoid, may well bring a new understanding to light concerning some particular aspect of flying which might just come in handy someday. While some of these may only extricate you from momentary lulls in conversation, others might come to mind someday when you need to pull out of a spiral dive, or in another case, decide between a downwind, uphill landing or one that is upwind but downhill. Whether you view flying as an art, or a science, here are some questions that invoke elements of both.

Subject: Like A Bat Out Of . . .

Question 1: The climb angles for most GA airplanes could hardly be described as spectacular. For most single-engine Cessnas out there, the actual climb path traveled is around six degrees nose-up from the horizontal (though the deck angle would be a bit greater than this). That's not exactly scalded ape territory. What would be the best climb angle for an Extra 300L ? (at Vy, STP, and an "Acro 1" weight--which is "light"--of 1808 lb)

A) 11 degrees
B) just over 19 degrees
C) exactly 45 degrees
D) straight up

Subject: Wing Buds:

Question 2: Air is thin stuff, but we can still fly in it. Just how small can the wings of a piloted airplane be, anyway? (excluding a NASA "lifting body", that is)

A) The officially recognized "world's smallest airplane" is the Bumble Bee, now on display at the Pima Air and Space Museum in Tucson, AZ. (There was also a similar airplane at the EAA museum in Oshkosh, WI.) It was (is) an experimental airplane built by an Arizona pilot. Its wingspan is only six feet, six inches!
B) Actually, the smallest airplane WAS the Bumble Bee II, built by the same person, which had a wingspan of FIVE feet, six inches. It was destroyed in a crash in 1988.
C) Antonin Przybylowicz (priz-BOL-oh-witz) of Buffalo, NY, flew his jet-powered experimental airplane, named Wingless Wonder, from the Mohave Desert in 1977. It had a takeoff speed of 220 kts, with a stall speed of 205 kts. It had NO wings, and obtained lift from entirely from

the fuselage, which had a wing-shaped cross section and a width of three feet, four inches.

D) The late actor Michael Dunn flew a custom-build airplane for the 1960's television show The Wild Wild West, which had a wingspan of eight and one half feet. This was not a dangerously small aircraft, however, considering the sole occupant, Mr. Dunn, was a dwarf. (He was a licensed pilot, and did not use a stunt man. Incidentally, he also had an IQ well over 160.)

Subject: Gee Whiz

Question 3: How can you make a level turn in an airplane without increasing the vertical g-load beyond 1.0?

A) The only way is to make a "great circle" turn. That means straight and level flight, where the turn radius becomes one half of the Earth's diameter (though not as some books might say, "infinite").
B) using rudder inputs only (and enough opposite aileron to counter any induced roll)
C) in a descending turn, where the rate of descent is slowly increasing
D) It isn't possible.

Subject: Looping the Loop, the Hard Way

Question 4: Who was the first person to perform (successfully: that is, and live) an outside loop?

A) Glenn Curtiss
B) Lincoln Beachey
C) Jimmy Doolittle
D) Junior Barnes

Subject: Highest Speed For A Jet

Question 5: In the vicinity of which altitude can you generally expect the highest true airspeed for a jet aircraft with a service ceiling of FL 370 (37,000 feet)?

A) FL 180
B) FL 250
C) FL 330
D) FL 370

Subject: Stop Da Music!

Question 6: In July 1990, on a very hot day, all operations stopped at Phoenix's Sky Harbor airport. It was a scorcher, but the density altitude increase of over 4000 feet wasn't the reason. What was?

A) MILSPEC requirements for an asphalt surface do not specify that they remain firm above a temperature of 165 degrees F. (Although the air temperature was over 120 degrees, the surface was much hotter.) Several airliners were stuck fast, like a modern day replay of the La Brea tar pits.
B) The air conditioning systems in all three terminal buildings overheated.
C) No airliner's performance charts went up to the 122 degree F temperature.
D) It was July 4, and the airport was closed.

Subject: "And for my next number..."

Question 7: Why would aerobatic stunt pilots do headstands during training?

A) For short periods afterwards, this increases oxygen flow to the brain, which helps prevent loss of consciousness during high-g maneuvers.
B) to increase the storage capacity of major blood vessels in the upper body, particularly the head, which helps prevent blackouts during high-g maneuvers
C) to acclimate the body to the upward surge of blood that occurs during negative g maneuvers
D) to gain proprioceptive familiarity with reversed outside references, or to put it less clinically, to get used to being upside-down

Subject: That Last Turn Is A Doozy...

Question 8: A fixed wing aircraft pilot, flying alone, enters a fairly wide canyon one morning--let's say it is a mile and a half wide--in his Cherokee at 100 kts, keeping at a comfortable distance (a half mile) from the right side of the canyon. He then executes a 30 degree banked turn to the left and reverses direction, clearing the cliff at the other side by an equal margin. Later in the afternoon, he brings three friends along, and repeats the maneuver. Wisely knowing that stall speed goes up with increased weight (or wing loading), he plays it safe and adds an extra 15 knots of airspeed to maintain the same wide margin above stall speed. (He's an engineer, and he knows that about matches the square root of the ratio of present to earlier gross weight.) Wanting to also compensate for increased stall speed by reducing his bank angle, as well as giving his right side passengers a better view with less discomfort, he turns with a more gentle 20 degree bank. What is wrong with this?

A) The true airspeed will also go up in the less dense afternoon air, and his safety margin against getting boxed in (and then boxed up) may vanish.

B) Weight actually has nothing at all to do with turn radius, but speeding up increases it. Turn radius varies with the square of velocity, so that 15% speed increase really means about a 32% increase in turn radius. That, plus sloppy technique could erode that half-mile safety zone down to 2000 feet or even less.
C) Turn radius varies inversely with the tangent of bank angle. That 10 degrees less bank actually means over a 50% increase in turn radius, so the safety margin might be 1500 feet or less, especially in the less dense afternoon air.
D) He's an engineer, but maybe not a very thorough one. Turn radius depends on both speed and bank angle. Going 15 knots faster and banking 10 degrees less actually results in over twice the turn radius. Since a 100 knot airplane in a 30 degree bank needs almost 3100 feet to turn around under ideal conditions, that half mile safety zone, especially in the warmer air, is now well under 1000 feet.

Subject: The Right Number

Question 9: We all know about best glide speed, but if you had an engine failure, were at a significant height right above your intended point of landing, and wanted to buy some more time, to what other speed might you trim up? (That's a hint.)

A) maneuvering speed
B) engine failure speed
C) levitation speed
D) minimum sink speed

Subject: Escaping Confinement

Question 10: To make the shortest possible takeoff, what other option is available to a pilot that does <u>not</u> involve changing the weight, control surface configuration, or applied power?

A) changing the CG
B) installing a climb prop
C) waiting for a good headwind
D) applying vaseline to all exterior surfaces

Subject: Differential Flight Dynamics? Vy/Vx

Question 11: True or false: V_y, the best rate of climb airspeed, decreases with altitude.

Subject: Here We Go Loop De Loop...

Question 12: A pilot lets himself get into a spiral dive, with bank and airspeed rapidly increasing. Coming to his senses near redline as the bank reaches 75 degrees, he quickly levels the wings. If he does nothing else, what happens next?

A) If the air isn't smooth, the wings may break off.
B) It will rapidly roll into a dive in the opposite direction
C) He will cause an accelerated stall
D) The airplane will enter a steep climb, and could even loop

Subject: Breaking the Bank

Question 13: A pilot is flying his Skyhawk and is practicing turns about a point. His airspeed is about 90 knots, his altitude is about 700 feet AGL, and his turn radius is about 800 feet. On the downwind side he finds it necessary to increase the bank angle slightly, but he is distracted and to his alarm, discovers that his bank angle has exceeded 60 degrees and is rapidly increasing. What is he in danger of doing? (More than one may apply.)

A) entering a graveyard spiral
B) rolling inverted
C) exceeding the design load of the wing
D) stalling if he increases back pressure to maintain altitude
E) descending to too low an altitude

Subject: Pushing the Limits

Question 14: The (positive) limit load factor for a normal category airplane is

A) 3.8, but only below V_{no}, the maximum structural cruising speed
B) 3.8 (or 4.4, if it is operated in the lighter "utility" category)
C) A and B
D) usually 3.8 (with the stipulation in A), but for certain normal and commuter aircraft, it could be as low as 2.2 !

Subject: The Cost of Speed

Question 15: What is currently the highest airspeed-to-horsepower ratio in a propeller airplane?

A) generally around one-to-one (knots per horsepower)
B) usually less than 1:1
C) two to one
D) almost three to one

Subject: Taking A Wrong Turn

Question 16: In a level turn, the number of "gees" a pilot pulls is, as we know, related to the angle of bank (actually it's the reciprocal of the cosine of the bank angle). Just as important as the increased load imposed upon the wings from the turn is the resulting increase in stall speed (which goes up by a factor of the square root of the reciprocal of the cosine of the angle of bank). However, there *is* actually one way that a pilot can yank and bank at will, and not impose any airframe stress or suffer any increase in stall speed. Just what is that?

A) By pitching downward at one third of the bank angle in a gliding turn, stall speed will not increase (although load factor still will).
B) There actually is no way to return the load factor and stall speed increase ratio back to unity. This is a trick question.
C) With a downward pitch of two-thirds the bank angle, both stall speed *and* load factor will remain unchanged from level flight.
D) By accelerating downward (steadily increasing downward pitch) while turning.

Subject: Stalling and Falling

Question 17: An airplane can be made to stall at any speed. Okay then, at what airspeed will *any* airplane stall, if the stick (or yoke) is briskly pushed forward so that it enters a rapid descent, and the airplane is "unloaded" to exactly zero gees?

A) By definition, it will immediately stall, regardless of the airspeed (even if it were zero).
B) At zero gees, stall speed is the square root of the cosine of the angle of incidence, multiplied by the quotient of present airspeed to the square of Vs. (Generally, since this works out to just a very few knots, the wings usually won't stall.)
C) At zero gees, an airplane cannot stall at any speed below Mach 1.
D) Mach, schmach...This is a trick question: The airplane won't stall, period. It can't, because at zero gees, the wing doesn't need to produce any lift!

Subject: Getting Pushy!

Question 18: Your average 172, at normal cruise speed, pushes about how much air downward?

A) In level flight, all air is propelled downward, and it must exactly equal the weight of the airplane.
B) Actually it is more meaningful to speak of a volume of air displaced rearward, rather than downward. Depending upon propeller design and installation, the general ballpark figure for engines of that size would be a static thrust of (very roughly) about 200 Kg. Since air at standard temperature has a mass of approximately one and a quarter kilograms per

cubic meter--we're talking mass here now, not pressure--that equates to about 160 cubic meters per second of air blasted backwards, each second.

C) an amount equal to about five times its own weight every second

D) As mentioned in choice B, most of the air is of course directed rearward. However there is a downward component which also equals the weight of the airplane. It is dependent on angle of attack, but it is time-independent.

Subject: Climbing On the Prop

Question 19: True or false: Apart from tilt-rotor or "powered lift" category aircraft, it is physically impossible for a propeller airplane (whether piston or turboprop) to either maintain altitude (or as the expression goes, "hang on its prop") *or* to climb straight up, by virtue of sheer thrust alone.

Subject: A Wing And A Prayer

Question 20: True or false: It is possible to maintain control of (as well as land) an airplane that is missing a significant part of a wing.

Subject: I'll Just Fly Between the Raindrops...

Question 21: What popular airplane had control problems when flying in rain?

A) the Wright Flyer
B) the Vari-EZ
C) the Spruce Goose
D) the popular Reno racer "Rare Bear"

Subject: Vector Analysis Isn't Just Tedious. Why, It Can Be Exasperating, Too!

Question 22: In a constant rate climb, the lift produced by an airplane's wing is

A) exactly equal to its weight
B) greater than its weight
C) less than its weight
D) equal to its weight, but only at L/D_{max}.

Subject: A Great Cross-Country, But He's Out Of Fuel...

Question 23: The greatest distance ever covered (either "straight line" or free) by a powered airplane (<u>or</u> for that matter, any sailplane) without benefit of a single drop of fuel was

A) in 1981, across the English Channel

B) in 1986, over the length of the Grand Canyon (about 165 nmi)
C) in 1990, across the entire United States
D) in 1992, from Vinon, France, to Fez, Morocco (about 747 nmi)

Subject: Making the Most Of It

Question 24: Which one of the following statements regarding best glide speed is incorrect?

A) With an engine failure in a retractable gear airplane, if the gear cannot be retracted, the airspeed should be increased to that above the published best glide speed.
B) When flaps are extended, best glide speed should be decreased.
C) In a headwind or a downdraft, glide speed should be increased.
D) At reduced weights, best glide speed should be decreased.

Subject: Magic In Numbers

Question 25: Which of the following statements is most correct?

A) For any airplane, the minimum sink and maximum endurance speeds are equal, and both are each about $^3/_4$ that of the best glide speed.
B) For any airplane, the best glide speed is also the speed for maximum range, both occur at L/D_{max}, and both are each in turn about 3/4 of the speed for optimum cruise (where you get the most "knots per gallon").
C) A stopped prop can increase glide ratio by up to 20%.
D) Altitude has absolutely no effect on the indicated speed to use for attaining an optimum glide, although best glide speed should be decreased at lighter weights.
E) Although one should increase glide speed by half the headwind component, if the gear remains extended in a retractable-gear aircraft, glide speed should actually be decreased by around ten to 15 percent.
F) All are correct.

Subject: Variable Sweepback

Question 26: Which aircraft was the first to have variable-sweep wings?

A) The Bell X-5
B) The F-111
C) The MiG-23
D) The Messerschmitt P.1101

Subject: On A Roll

Question 27: To the nearest 100 feet, estimate the takeoff distance into a 13-knot headwind for an airplane with a 52-knot takeoff speed where, based on current conditions of temperature, pressure, runway surface, slope, etc., the no-wind takeoff distance is given as being 1000 feet. Then estimate that distance with a 13-knot tailwind.

A) 1000, 1000
B) 800, 1200
C) 600, 1600
D) 400, 1900

Subject: Good News, Bad News: Uphill Headwinds

Question 28: You and your Cherokee are at a 2500-foot runway having a significant slope, where the elevation of the high end of the runway is 1900 feet, and at the other end, it's 1812. It's a nice, mild day in the early spring, and the temperature is about 60 degrees. The wind speed is 15 knots, right down the runway. The trouble is, it's really *down* the *runway*! Which would be best, taking off downhill, but with a hefty tailwind, or uphill, into the wind? There are 50 foot trees at either end (of course). How do you decide which way to take off?

A) You don't. Wait until evening, when the wind dies down, and take off downhill.
B) Take off downhill, being sure to configure the airplane for a short-field takeoff.
C) Take off uphill.
D) There isn't enough information to answer this question.

Subject: Noises Off

Question 29: You're descending at a relatively high speed in calm air, and since there was no forecast for nor little chance of turbulence, you let the airspeed climb into the yellow arc, but fail to keep your attention focused on the airspeed indicator. Then you hear this buzzing noise. What might that mean?

A) You probably had a bee, wasp, or other insect as a stowaway.
B) Your mike is stuck.
C) One or more Dzus fasteners have come loose, most likely on the cowling.
D) You'd better slow down!

Subject: Tales From the Crypt

Question 30: What is the "coffin corner"?

A) the dead end of an uphill one-way Idaho back-country airstrip

B) a narrow airspeed range, within which one can be on the verge of flying both too fast and too slow, at the same time
C) a politically incorrect term for the NTSB reference library
D) a region formerly part of Hungary, now Romania, above which many battles were waged between Soviet and eastern European fighters during WW II

Subject: Mach 1, Standing Still

Question 31: When traveling at the speed of sound at sea level (assuming we use a "standard" atmosphere at 15 degrees C), the dynamic pressure (the same pressure sensed by a pitot tube) is very roughly equivalent to

A) one fifth normal atmospheric pressure at sea level
B) normal atmospheric pressure at sea level
C) five times normal atmospheric pressure at sea level
D) fifteen times normal atmospheric pressure at sea level

Subject: Blowing In the Wind

Question 32: The first person to propose the use of wind tunnels to gather aerodynamic data was:

A) Francis Wenham, in 1871
B) Samuel P. Langley in 1891
C) Wilbur Wright in 1901
D) Edme Mariotte in 1686

Subject: Different Meanings

Question 33: In an airship, what does the term "pressure height" mean?

A) when the cells become 100% full
B) when the airship flies through a low pressure area
C) when the airship deflates due to very high pressure surrounding it
D) when the strain on the envelope is within 10% of burst pressure

Subject: An E-ticket Ride You *Would* Want To Miss

Question 34: On April 4, 1979, TWA Flight 841, a Boeing 727-100, registered N840TW, prepared for departure from John F. Kennedy (JFK), New York, for a scheduled service to Minneapolis / St. Paul International Airport in Minnesota. During cruise, over Saginaw, Michigan at flight level 390, the 727 suddenly plunged towards the ground at an average descent speed of

A) 10,000 feet per minute

B) 15,000 feet per minute
C) 20,000 feet per minute
D) *more* than 40,000 feet per minute

Airmanship and Aerodynamics: Answers

Answer to Question #1: The answer is choice B. In the 300, the best rate of climb is 3200 fpm, though it's lower at MGW, and Vy is 96 knots. (Though Vx is much lower, at 87 knots, the climb rate at Vx is not published.) Almost no present-day piston powered production airplane can reach 45 degrees, though there were some experimental military airplanes, the "tail-sitters", that actually could climb vertically, "on the prop". (The landings were, shall we say, dicey.)

Answer to Question #2: It's choice B again. The builder, Mr. Robert H. Starr, of Phoenix, Arizona, was also the pilot. (He completely recovered from his injuries.) The Bumble Bee II cruised at 150 mph, had a top speed of 190, a stall speed of about 86 mph, and an empty weight of 396 lb. Choices C and D were alas almost complete fabrication. (Michael Dunn's IQ was really that high, though.)

Answer to Question #3: Allowable answers are choices B or C.

Answer to Question #4: The correct answer is choice C. On May 27, 1927, United States Army Air Corps Lt. James Doolittle flew the first outside loop over McCook Field near Dayton, Ohio in a Curtiss P-1B Hawk. Scientists had predicted that outside loops (inverted loops) would be fatal. (They usually were, as were many attempted afterwards. It ain't easy.)

Answer to Question #5: It's choice B. Jet aircraft typically have two speed limiters, indicated air speed and Mach speed. The highest available true airspeed occurs where the indicated speed curve, a curve that yields higher true airspeed with an increase in altitude, intersects the Mach speed curve, a curve that yields a reduced true with an increase in altitude. There can be a significant variance however, depending on aircraft type. Vmo can either stay the same or increase slightly with altitude, and the maximum true airspeed occurs at an altitude where Vmo is greatest (before being reduced because of Mmo). Actually, Mach speed is primarily dependent on temperature. There is also one other slight dependency (humidity), but it is so small that it is effectively ignored in Mach speed calculations.

Answer to Question #6: It was C.

Answer to Question #7: C is the one. Aerobatic pilots must deal with negative as well as positive g's. The familiar positive g's occur when flying maneuvers like a normal loop, where blood drains from the head and pools in the feet. Some pilots lose color vision or acuity, since the eyes are very sensitive to oxygen level. For others, vision stays sharp, but narrows, and although their thinking is not impaired, one minute they're marveling at their sudden tunnel vision, and then it's lights out. (This is G-LOC, or "g-induced loss of consciousness.") Uncle Sam provides "anti-g suits" to its pilots, but flying "outside" loops poses a reverse challenge, with blood now rushing to the head, risking nose bleeds or burst vessels. Negative-g's are more difficult to nullify. (You can't strap a g-suit on your head.) So one preventive measure, as funny as it looks, is for pilots to do headstands and handstands to condition themselves for that head-ward blood surge (or go to the nearest tot lot and hang from the monkey bars). Can't say that I've ever seen it, though...

Answer to Question #8: It's choice D. That would be quite a pucker factor for those passengers as he came around towards the other side of the canyon. The buzzards are waiting...

Answer to Question #9: The answer is D; minimum sink speed, which is generally about halfway between Vx and stalling speed. You wouldn't glide as far, but you would sink more slowly in the vertical.

Answer to Question #10: It's choice A. Taking all onboard items and moving them to the rear-most allowable position (still remaining within aft CG limits) will reduce stall speed, liftoff speed, and takeoff distance. (Think of it as helping provide a tail-down force without exacting as great of an aerodynamic penalty.) Choices B & C would work as well...

Answer to Question #11: FALSE. The true airspeeds for both Vy and Vx actually increase with altitude. However, because Vy increases only slightly, the indicated best Vy speed usually does decrease with altitude. The indicated Vx speed however actually increases.

Answer to Question #12: It's choice D. Both the FAA Flight Training Handbook and the Instrument Flying Handbook neglect to mention that upon recovery you actually need to push nose down (in addition to retarding the throttle of course) because if you do not, you may find yourself on your back in IMC (or at least as bad, in a tail slide), having just escaped the spiral. After being in the frying pan of an impending dive, the natural reaction is to pull up after leveling the wings. However, there's one small problem: that's the airplane's reaction, as well! Even a fairly light and relatively boxy airplane like a Citabria will enter at least a 60 degree nose up attitude, which is exactly what Barry Schiff experienced when he tried it for me. Conceivably, it is possible that an aerodynamically cleaner aircraft could go farther, and possibly "over the top". If you pulled back in addition to the airplane's own tendency to generate much greater lift at the airspeed you would likely have reached by this point in a spiral, almost any airplane could easily enter a loop, or (like I said) possibly worse, a tail slide.

Answer to Question #13: choices D and E: Although most 172s do not have enough power to maintain level turns beyond about 55 degrees of bank, stall speed still increases greatly, and so will altitude loss, especially if he stalls the wing and enters a spin. The airspeed is not high, so the wings will stall before reaching load limits. (Hopefully he's in VMC, so the first two choices don't apply!)

Answer to Question #14: Choice D. Choice A is true, because that load factor goes from 3.8 to just one between maximum structural cruising speed and never-exceed speed. And choice B is true for many normal category aircraft, but not all. As far as the surprising statement in choice D goes, for heavier aircraft (potentially, any aircraft grossing over 4117 pounds) Title 14 CFR Part 23.337 actually states that the positive limit maneuvering load factor for normal and commuter aircraft may not be less than 2.1 + [24,000/(W+10,000)]. Conceivably if there were a normal category airplane that weighed 300,000 pounds, the load factor might be +2.2 (though actually most transport category airplanes are limited to +2.5). It also states that the positive load factor need not be more than 3.8, and the negative load factor may not be less than 0.4 times the positive load factor for normal, utility, and commuter categories. (For the utility category, the positive load factor of course is 4.4; for aerobatic airplanes, it's 6.0; and the "negative" coefficient for aerobatic airplanes becomes 0.5, not 0.4.)

Answer to Question #15: Choice D. The speed of an airplane is limited by the friction it encounters with the air through which it flies. For airplanes not powered by jet engines, the efficiency of a propeller in converting horsepower to thrust is reduced as airspeed increases. Although one form of drag called induced drag actually decreases with increasing airspeed, generally overall drag increases in proportion to the cube of the airspeed, and so airspeed increases by only the cube root of the ratio of increased to original horsepower (though the climb rate is linearly related to excess horsepower). Most fixed-gear general aviation airplanes generally get somewhere in the vicinity of one (and often less) mile per hour for each horsepower that its engine is capable of producing. In aviation, as well as with ships at sea, the preferred unit of speed is the knot, which is equivalent to about 1.15 miles per hour, but this "one-to-one" ballpark figure--plus or minus around 50%--is still generally true for most small general aviation airplanes. For example, a new Cessna 172, with a 160 horsepower engine, has a maximum sea level cruise speed of 123 knots, or around 141 mph. But there are some well-known exceptions.

Incredibly, it is about three-to-one. For example, an owner-designed and built airplane powered only by a 65-hp Rotax engine has achieved a speed of 213 mph in 1992. Having the very lightest FAI land plane weight class (C1a/0, under 300 Kg. gross weight), the speed-to-power ratio of Californian Mike Arnold's AR-5 was about 2.85 knots/HP (3.3 mph per hp). Where most small GA airplanes have an equivalent "flat plate area" (not literally a frontal cross-sectional area, but still a useful metric) somewhere around three to six square feet, this airplane has an FPA of well under one square foot. Austrian Peter Scheichenberger's Bede BD-5B achieved over 218 mph (almost 190 knots) in August 1999, so as far as FAI records go it was faster, but that was with a 74 HP Rotax (and a "knot/HP" ratio of 2.56).

Mike Arnold's AR-5 and Peter Scheichenberger's Bede BD-5B

Answer to Question #16: choice D. Most people would consider it intuitively obvious that stall speed will not increase during a gliding turn, as it will during a level one. Actually though, during a stable gliding turn (that is, at a constant sink rate), the effect upon stall speed is *exactly the same as it is in level flight*! To prevent the stall speed from increasing, one must unload the wings. A way to do that would be by accelerating downward while turning. If one used a calibrated G-meter and matched the negative gees to one's previous load factor, it could probably be matched fairly closely. As long as sink rate is increasing, stall speeds will fall as well. (Eventually though, you might run out of altitude.) And as you stop the drop and sink rate is reduced back to level flight (i.e., zero), stall speed would again rise according to whatever bank angle you still had. But since few of us fly with a g-meter, and negative gees are hard for most of us to judge accurately, this little physics factoid is best approached with caution!

Answer to Question #17: It’s choice D. Don't you go telling me that I got you with *this* one!

Answer to Question #18: The right one is C. The other verbiage is mostly chaff. Bulls-eye if you guessed the right answer. It is indeed true at least in a very rough sense that a propeller (one of perfect design) might achieve four pounds of thrust for each available horsepower, but wings keep us up by pushing air *down*. That Cessna will be pushing about five times its own weight of air down, each second. Air at standard temperature "weighs" (really, it's mass and not weight) about 1.25 ounces per cubic foot or also, coincidentally, about 1.25 Kg per cubic meter. (The more precise numbers are about 1.22 and 1.29, respectively.) If our Skyhawk weighs one ton, that would be about 4200 cubic meters of air every second. Visualized as a cube of air, that would be a box that's a bit over 16 meters on a side (or not quite 53 feet) every second!

Answer to Question #19: False. Obviously, we have all heard of aircraft such as the Bell XV-15, the V-22 Osprey, or the Bell 609. *These* don't count. And some jet fighters such as the F-16, with a thrust-to-weight ratio greater than one, can accelerate going straight up. And of course one might think of the Harrier Jump Jet. But *those* don't count, either. Oh, and zooming upward, trading kinetic energy for the potential energy of altitude--well, that's obviously cheating. We're talking raw, blistering, prop fury here.

It can be done, and it has been. First, a little background. For a helicopter or anything else going straight up without the assistance of a rocket, the force of lift is proportional to the mass of the air directed downward, as well as its speed. You can accelerate a lot of air at a modest velocity, or a small amount at a tremendous velocity, or anything in between. It takes power to do that however, and how much depends on the kinetic energy imparted, according to that old equation:

$$KE = {}^{1}/_{2}\, mv^{2}.$$

This tells us that the least expensive way to shove the air down is to push on gobs of it, as slowly as we can get away with. Well, propellers aren't too great at that job. But curious minds want to know, and gosh darn it, we built *two* different models. (This was in the early 1950's, of course.) Intended as experimental VTOL fighter aircraft, there was the Lockheed XFV-1, and there was the Convair XFY-1. Both had maximum horizontal speeds around 500 mph and both had monstrous turboprop engines driving coaxial contra-rotating propellers. They sat on cruciform tails having small castoring wheels (without brakes) at their tips, and their respective pilots each sat on gymbaled seats. They were known as the "tail-sitters" or in the case of Convair, the "pogo". Lockheed's entry never made takeoffs and landings from the vertical, but the XFY-1 did. For the precise reason that their very flight regime epitomized the word "precarious", these birds got the axe after only a few months.

the Lockheed XFV-1, and at right, the Convair XFY-1

Answer to Question #20: It's true. In March 1945, a pilot was flying the military version of the DC-3, a C-47, when his left wing was severed in a mid-air collision. The wing, from just past its left engine out to its wingtip, was completely gone (or as the FAA might say, it had departed the aircraft). The wingspan of a C-47 was 95 feet, six inches, and the distance from just past an engine to a wingtip was about 36 feet. So that would mean it lost about 38% of the total wingtip-to-wingtip wingspan, or about 42% of its wing area! He managed to maintain control and land (sort of). In another case, a North American B-25 Mitchell bomber, from the 77th Bomb Squadron in the western Aleutians, was making bomb runs on the northern Japanese islands during the end of World War II (also 1945) when ground fire took out a large portion of the left outboard wing panel, including the aileron. They also managed to limp back to their base on Attu Island, in the Aleutians. Other similar incidents have occurred, some not quite as severe, such as when a commercial DC-3 pilot took his passengers sight seeing near Phoenix, AZ in 1957. He got too close to a mountain and took a few feet off of the left wing tip, but he made a normal landing. In another instance, a mid-air collision clipped 5 feet from one wing of a Capital Airlines DC-3, but its pilot also landed safely. There was also the Pan Am Boeing 707 that departed SFO in June of 1965: The number four engine disintegrated, taking 25 feet of the right wing with it. For the B-707, the wingspan was 146 feet, and 25 feet off one wing would be about 17% of the wingspan, or with a cabin width of roughly 12 feet, would be equivalent to losing about one-fifth of its total wing area. Nevertheless, an emergency landing was safely made at Travis Air Force Base, with no casualties. According to Barry Schiff, all you need is "enough wing on the mishap side to prevent a roll." In the case of the B707, it had manually operated flight controls (except for the rudder, which was hydraulically controlled). But the biggest difference between transport category aircraft like the 707 and light airplanes is that the cables to the ailerons (and in the case of the 707, the spoilers) were independent of one another, so losing the cables to the outboard aileron wouldn't affect everything else.

So how does that translate down the food chain to the pilots flying those smaller general aviation airplanes? On almost any airplane, the pilot's control inputs are transferred out to the ailerons either by a system of pushrods or via cables and pulleys. On a Cessna, the cables for each respective wing are not all on one "clothesline" type arrangement between two pulleys on either end, and there *is* some independence between the two. However, on almost any airplane, there are articulated mechanisms, such as bellcranks, usually located on the inboard end of each aileron, for transferring control inputs directly to these very crucial surfaces. If one were to become jammed by torn or bent metal however, that could conceivably lock up the controls for *both* ailerons. This would apply to the pushrod type systems, as well as those with cables.

Answer to Question #21: choice B. Actually, both the Vari-EZ and the Long-EZ, which have laminar flow canards, experience aerodynamic changes and exhibit some significantly adverse changes in control characteristics when flown in rain. This is because they are laminar-flow wings, and when it rains, the raindrops trigger boundary-layer turbulence. Likewise, deciding to paint decorative stripes on the leading edge of a canard can--and has--also caused difficulty in maintaining pitch. So might flying through a swarm of bugs. (yuck!) Laminar flow wings actually stall at a lower angle of attack, as the air does not stay "attached" to the wing as long as when the boundary layer air is churned up, which is one reason that some pilots install vortex generators to delay the onset of a stall!

Answer to Question #22: It's choice C. In any sustained climb, the lift produced by the wing is actually less than the weight of the airplane. How can this be so? In unaccelerated straight and level flight of course, lift equals weight. Now take an F-16, climbing at a steady rate, straight up. The wings are contributing nothing; that GE or Pratt and Whitney is doing all the work. Okay, now take your Beetle Bomb Bug Smasher, climbing like a scalded ape at a dizzying climb angle of, oh, six degrees. Does the same thing hold true? Yup. When you pull back on the controls, whatever thrust is coming from that prop, multiplied by the sine of your climb angle (which would work out here to about one tenth of it) is what's helping hoist you skywards. (So now only nine-tenths of that thrust is left to counteract drag forces, and guess what? You slow down.)

Answer to Question #23: Try C. MacCready's Solar Challenger first flew several miles in Arizona in the late 1970's before its historic flight across the Channel in 1981. Actually, it didn't just hop the 21 or so miles across the narrowest portion of the Channel from Dover to Calais, but instead flew about 163 miles from Corneille-en-Verin Airport north of Paris, to Manston Royal Air Force Base south of London. However, in 1990, an aircraft named Sunseeker completed a 2,523 mile trip from Desert Center CA to Spot, NC, and became the first airplane to cross the United States almost exclusively by solar power. (I said "almost" because it did use lift from thermals as much as possible.) At any rate, it was certainly the longest "fuel-less" flight ever attempted. It took the pilot, Eric Raymond, a bit ovr 125 hours to make the flight. Actually that's flights, plural (21 of them). He left California on July 16, and due to things such as adverse weather, mechanical problems (and the need for an occasional charge) he and his 198-pound graphite fiber and epoxy airplane, which was powered by a 3 HP wheelchair motor driven by electricity from about 88 square feet of solar cells, didn't arrive on the East coast until September 3rd.

The average length of the intervals during which the motor was used was only five minutes, and these were mostly takeoffs and traveling from one thermal to the next. For those double-E majors out there, the PV array produced about 2.5 amps at 120 volts, and charge time for the batteries was about two and a half hours. The no-fuel flight over the Grand Canyon is total fiction, and the flight in choice D happens to be the current FAI "straight distance" record for any sailplane, set on April 17, 1992. (The longest glider flight ever, incidentally, occurred on December 12, 2002, and was a "free three-turnpoint" flight over western Argentina, of about 1630 statute miles.)

the electric powered glider Sunseeker, ca. 1990
photo reproduced with permission from Eric Raymond of Solar Flight Inc.

Answer to Question #24: The right "wrong" one is A. All aircraft with wings have a best glide speed. So for that matter do certain flying fish, reptiles, birds, and mammals (bats). Some aspects of aerodynamics behind it are however less than intuitive. In this case, best glide should actually be decreased with the gear extended. To be precise, best glide speed actually decreases by the fourth root of the effective increase in "equivalent flat plate area". Doubling it (which is about what happens extending the gear in a Bonanza) would require decreasing best glide by about 15%. You would also unfortunately glide more steeply and not as far, but that would become your new optimum glide profile, as well. It might seem odd that in order to maintain an optimum glide profile in a headwind, one must speed up (choice C), yet if you "make your own headwind" by adding drag, you must instead slow down. Slowing below the normal best glide speed also holds true for adding full flaps (choice B). As a rule of thumb, when flying against a headwind or when in a downdraft, one should increase glide speed by about half the headwind component or about four to five knots per hundred feet per minute of sink rate. (For tailwinds and updrafts, that would be about one quarter of the tailwind component, and two to three knots per 100 fpm, respectively.) Best glide speed also decreases proportionally with the square root of weight.

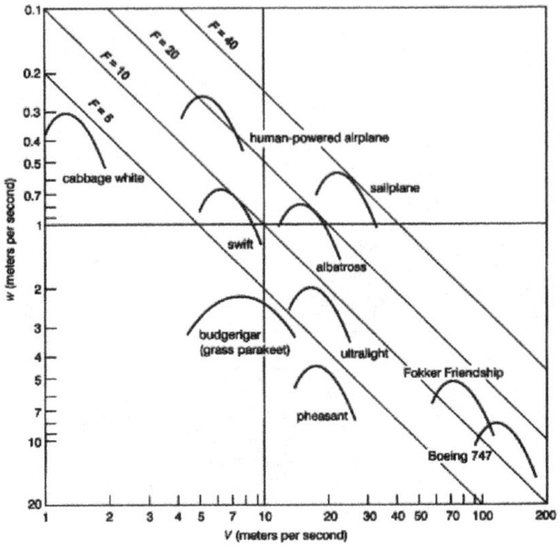

a speed polar graph (published with permission) from the Danish sailplane web site
http://hjem.get2net.dk/helmuth/data-diff.html

Answer to Question #25: In the spirit of conveying as much truth and knowledge as succinctly as possible, the best answer is in fact choice F. The exact ratios in choices A and B are actually the fourth root of one third (not quite 0.76). For choice C, there is no one exact number for every airplane. For choice D, best glide speed goes down with the square root of weight. For choice E, best glide decreases with the fourth root of "effective flat plate area" (which is only about 15%, when this value is doubled...which is close to what happens when the gear is extended, for example in a Bonanza). But of course the most remarkable thing concerns the speeds for maximum range, maximum endurance, and minimum sink, as well as optimum cruise. They can be very useful indeed, and although they rarely appear in an airplane owners manual or a POH, they can however be quickly and easily calculated based upon a number that usually *is* given, namely, the best glide speed.

Answer to Question #26: D. The P.1101 was to be a test bed aircraft used to study the effects of swept back wings on high speed flight but was not completed before the end of WWII. As early as 1935, German aeronautical engineers began studying the effects of swept back wings on high speed flight. Among them was the noted Alexander Lippisch, who filed a patent on a variable sweep wing in 1942. Dr. Waldemar Voigt, Lippisch, and others started designing a single-seat fighter interceptor, known as Project P.1101, backed by Messerschmitt. Due to extensive commitments to the Me 262, work on the 1101 went slowly. That is, until 1944, when the Luftwaffe commander issued a requirement for a single-seat, single engine, turbojet fighter with a higher speed than the Me 262. Messerschmitt entered the P.1101 design, got the contract, and started with design and production of a prototype at Oberammergau. The wings could be set to two different positions, forward position, with a relatively small 10-degree sweepback, and a rearward position with wings at a much sharper angle. However, they could only be moved *while the aircraft was not in flight.* Sometime either before or during WWII, someone figured out that the critical Mach number (or whatever they might have called it then) was related to the sweepback angle (actually, its cosine).

Before the prototype could be completed, however, US forces discovered the Oberammergau facility and under supervision of the chief engineer at Bell Aircraft, the P.1101 was transported to the USA, where Bell designed a better mechanism for adjusting the sweep of the wings, *in flight.* In 1949 a variable sweep wing aircraft, the Bell X-5, was designed based on the P 1101. The Bell X-5 made its first flight on June 20, 1951, and became the first jet-propelled aircraft ever to fly with in-flight variable-swept-wing capability, and formed the basis of modern variable-sweep wing aircraft. It was designed to sweep its wings from 20 degrees, to 45, and to 60, to see what Mach drag did on variable swept wings.

The Bell X-5 first flew on June 20, 1951. The F-111 "Raven", which first flew on December 21, 1964, had angles from 16 degrees (full forward) to 72.5 degrees (full aft). The MiG-23 "Flogger", which first flew in April 1967, had variable-sweep wings at 19, 48, or 75 degrees.

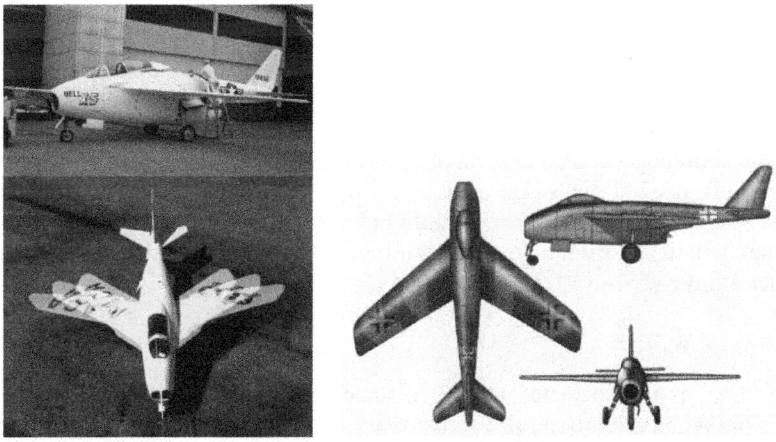

Bell X-5 P.1101

The Bell X-5, and illustrations of Messerschmitt P.1101

F-111 (top) and MiG (bottom)

Answer to Question #27: The given takeoff speed and takeoff roll were hypothetical numbers, but they are more or less representative of ballpark figures for a Cessna 172. For starters, are the numbers given in choice A realistic? No, and you can relax; choice A is laughably out of the running; I threw that in just to start you thinking. Okay: you probably noticed that 13 is exactly one quarter of 52. First off, a wind that's one quarter of your takeoff speed cuts the takeoff roll almost by half, and in the case of a tailwind, increases it again, by *more* than half. The mathematical formula for determining the effect of a headwind on takeoff distance is given by

$Dw = D_o(1 - W/V_r)^2$

where "Dw" is the corrected takeoff distance, "Do" is the no-wind takeoff roll, "W" is the wind velocity and "Vr" is the takeoff airspeed. In words, you take the headwind divided by the takeoff speed, subtract that quotient from one, square the result, and multiply *that* in turn by the distance for a no-wind takeoff roll. The "tailwind effect" formula is similar, differing only by there being a plus sign instead of a minus sign in the parentheses. It becomes

$Dw = D_o(1 + W/V_r)^2$.

Note incidentally that this is not a *precise* formula; the more intuitively obvious part of the algorithm (adding or subtracting the ratio of wind to takeoff speed to or *from* one, and multiplying some exponent of it by the no-wind takeoff roll) is one thing. The exponent itself isn't. As a matter of fact, the exponent used by the Flight Test Guide for Certification of Part 23 Airplanes (AC 23-8B in a recent iteration) is **1.85** !

Anyway, for a 13 knot headwind, you'd get
$1{,}000 \times (1 - 13/52)^2 = 1000 \times (1 - 1/4)^2 = 1000 \times (3/4)^2 = 1000 \times 9/16 = 9{,}000/16 = 563$ ft.
For a 13-knot tailwind, that's
$1{,}000 \times (1 + 13/52)^2 = 1000 \times (1 + 1/4)^2 = 1000 \times (5/4)^2 = 1000 \times 25/16 = 25{,}000/16 = 1563$ ft.

So the right answer is choice C.

Of course, you've probably heard of the even-simpler rule of thumb that a 10% headwind (that is, 10% of liftoff speed) reduces the ground roll by 19%, and a 10% tailwind increases the ground roll by 21%. This rule is exactly correct in fact, but only when the wind is exactly 10% of the takeoff speed (and of course, when there *is* no wind). Using multiple proportions at higher wind speeds (say by doubling that 19% when the headwind is 20% of liftoff speed) introduces errors--small at first, then progressively greater. The problem is that this rule of thumb approximation yields takeoff distances which are increasingly optimistic *under*estimations of what you would *really* need. For instance, with a headwind of 20% (in this case, 10.4 knots) you would come up with a takeoff roll only three percent shorter than if you had used the formula (620 feet, versus 640 feet). With a 30% headwind (15.6 knots) your optimistic underestimation represents a 12% error (430 vs. 490 feet). And with a 40% headwind, or 20.8 knots (by which point, in reality, you would probably be questioning your ability to taxi to the runway, let alone take off), this error becomes 33% (240 vs. 360 feet). Similar errors exist for the tailwind side of things, but the error (again, optimistic, or shorter than you would actually need) is much less: about 1% at 10 knots, 4% at 16 knots, and 6% at 20 knots (and very academic, since no one in their right mind would really take off the wrong way, with a wind that strong). Incidentally, for those mathematically inclined who might be wondering, the decrease in takeoff roll goes down by "X" times when headwind is exactly (1 - 1/SQRT(X)) of the takeoff speed, and the *in*crease in runway needed goes *up* by "X" times when headwind is (SQRT(X) - 1) of the takeoff speed. Also, the formula for decrease and increase in takeoff *times* (taking off against and with a wind, respectively) are similar to the distance formulas, except there is no "square" exponent.

Answer to Question #28: You may have asked yourself this question, somewhere along the way in your flight training (or afterwards): Is it better to take off uphill with a headwind, or downhill with a tailwind? (And even before being in *this* potential pickle, is it better to *land* downhill into a headwind, or land uphill, with a tailwind?) Just like the previous question regarding headwinds and tailwinds, there are other rules of thumb. One says that for down-sloping runways, you decrease takeoff distance by 5% for each degree. Another says that for up-sloping runways, every 1.0% grade has the same effect as an approximate 10% decrease in runway length. Yes, it's confusing. And yes, there are many other variables involved. If you're taking off downhill from a runway that's only 1000 feet long, but the other end is 50 feet lower, then it's like you had about 1150 feet. Of course, it gets complicated if the wind is also blowing downhill, too! Add in all the other "real world" factors, and this academic exercise can get pretty useless. It would also depend on the location of any terrain (particularly *rising* terrain) along your departure path. And we're assuming the runway has the same slope along its entire length (no bumps or dips...like you'd probably find in the real world). Let me elaborate on this. (It's worth the time.)

Any runway with a slope greater than or equal to just one third of one percent (actually, it's 0.3 percent) is noted in the Airport/Facility Directory (with the direction of the slope, when it's available), and on Instrument Approach Procedure (IAP) charts. First, on landing downwind: As already noted, a 10% increase in groundspeed gives a 21% increase in landing distance; even slight tailwinds noticeably increase the landing roll. If that runway is on a mesa or along a river, and you're landing uphill with a tailwind, you could get an updraft on short final, and you could find yourself floating further down the runway before you can get it down. Also, when you land with a tailwind, you have to effectively fly steeper approaches to compensate for your increase in groundspeed, and this can cause illusions that can make it harder to judge your position relative to the runway. Then, as you've no doubt also figured out, *taking off* with a tailwind also results in a flatter climb-out, making it harder to out-climb rising terrain. As far as landing downhill into a headwind, first of all the wind is going to need to be strong to cancel out that downhill slope, right? That can mean more turbulence on the approach, especially with uneven terrain. You might wind up having to fly the approach faster to compensate for gusts, and the increase in your groundspeed will of course increase your roll-out. Also, landing downhill, unless you have done it many times before, your airplane will probably float forever, because the ground will seem to keep dropping out from under you! Once you *are* on the ground, I hope your brakes work really well; you'll have an effectively heavier airplane, which might be rough work getting stopped in time.

There is actually a formula to determine the right thing to do. (And again, there's still another old rule of thumb: This one says that for most small airplanes, it generally will take about a 1.0% uphill slope to counteract the adverse effects of each two to three knots of tailwind--or, put another way, you'd need two to three knots of headwind for each one percent of a downward sloping runway.) There are actually formulas that address this, quantitatively. There is a very complicated one by Herrington, et. al. (*Flight Test Engineering Handbook*, AF Technical Report 6273, Air Force Flight Test Center, Edwards AFB, 1966.) Here is a simpler one, although it is still an approximation, by John Lowry, a noted aviation physicist and author of many articles (as well as a book) on airplane performance. The formula quantifies a wind velocity, which he calls the "break-even wind speed" which is the wind velocity that would give you the same distance to lift off in, whether you went uphill or downhill. Here it is:

$$V_{break\text{-}even} = \frac{\theta \times d_{lo}}{5 \times V_{lo}}$$

The angle theta is the runway up-slope (in degrees); d_{lo} is the POH "no wind" takeoff distance on a level runway (that is, for whatever the conditions are at the time, of course); and V_{lo} is your lift-off speed in KTAS. This runway has an 88-foot drop over 2500 feet, which means...let's see...the arctangent of 88/2500...that's a tad over two degrees. (Something to tuck away in the back of your mind: a 35-foot elevation difference between opposite ends of a 2000-foot runway gives just about one degree of slope. A one-in-a-hundred slope, say a 21 foot difference for a 2100-foot runway, is about 0.57 degree.) Let's say that your liftoff speed is 55 knots, and your POH says that you would need 1200 feet at whatever the particular density altitude was, at that moment. Plugging all that in says that the break-even wind speed is about nine knots. Since the actual wind is well over that, taking off uphill to take advantage of the wind is actually the right thing to do. So...after all this: the right answer is choice C.

If you're still awake, here's one more minutiae of interest. The Flight Test Guide for Certification of Part 23 Airplanes (AC 23-8B), mentioned in the previous explanation) gives the following equation to relate the effect of runway slopes:

Runway Slope. The effect of runway gradient can be significant for heavy airplanes or for low thrust-to-weight ratio airplanes, even if the gradient of the runway is small. Gradient should be controlled by proper runway selection. The correction is as follows:

$$S_G = \frac{S_{GS1}}{1 \pm \left(\frac{2gS_{GS1}}{V_{TO}}\right)\sin\theta}$$

Where:

	S_G	=	no-wind takeoff ground distance (feet)
	S_{GS1}	=	ground distance on a sloping runway
	g	=	acceleration of gravity, 32.17 ft./sec^2
	V_{TO}	=	airplane velocity at lift-off in ft./sec. (true)
	?	=	angle of the slope in degrees (not percent)

\+ for upslope and - for downslope

Answer to Question #29: Although aileron buzz usually exists in transonic airplanes, and is generally not an automatic precursor to flutter, it can precede structural failure in even an ultralight aircraft. True, the ailerons of certificated light planes are not supposed to exhibit aileron buzz even at never-exceed speed (red line), but it can--and has--happened. The reason can be an imbalance condition or problem in the control cables, for example. The "buzz" usually refers to a rapid oscillation of a control surface, most often an aileron, which doesn't usually attain a large amplitude or become dangerous, and is caused in most cases by shock-induced separation of the boundary layer at high transonic speeds. Nonetheless, you have just become a test pilot, and the only correct thing to do is to slow down (choice D). Again, such events have usually occurred with high speed operations and at high altitudes.

Answer to Question #30: It has nothing to do with overestimating the influence of gravity against brevity, sarcastic barbs against morbid preoccupation, or Transylvanian transgressions. It's about speed. As we know, an airplane's indicated airspeed decreases as its altitude increases, due to the greatly reduced air density, while its actual or true airspeed actually increases. Because the air is much less dense, the angle of attack has to be increased to maintain the same lift, but because of that and the already high true airspeed, the air must become more greatly accelerated over the wing, attaining speeds near Mach 1, which induces flow separation and buffeting. Eventually, the gap between a "low speed" and a "high speed" or Mach related stall continues to narrow, the higher you go, and finally stall speed and high speed Mach buffet merge in a way, and any sudden roll, increase in angle of attack, load factor, gusts, or turbulence could result in a loss of control. Theoretically, according to Barry Schiff, an airplane like a U-2 in a turn at high altitude could experience a stall buffet on the inside wing, and at the same time, Mach buffet on the outside one. Manifestations often include more typically low speed problems such as Dutch roll (coupled oscillations of roll and yaw), adverse yaw, and stalling. Undoubtedly attracted by both the potential alliteration and the opportunity to wax sepulchral, aeronautical engineers didn't miss their chance to label this region the "coffin corner". The answer is B.

Answer to Question #31: As anyone who has stuck their arm out the window of a moving car knows, moving air exerts a force. For those of us (the majority, I suspect) who have at least once in our youth stuck our outstretched palm out of a car window broadside at highway speeds, we have discovered that it exerts quite a lot. This force, when expressed per unit area, equates to what we know as pressure. In the case of moving air, or moving *through* the air, this is called *dynamic* pressure. Often symbolized by a variable using the letter "q", dynamic pressure varies according to the density of the fluid in question as well as the square of an object's speed through it. An equation quantifying this can be written as

$q = \frac{1}{2} r V^2$

where

r = air density

(in Kg/cubic meter, or slugs per cubic foot in our own quaint system), and

V = air velocity

(in meters per second, or feet per second)

As an aside, note that the units of dynamic pressure resulting from the product of these terms might not be what you would expect to see, such as pounds per square inch (or foot, in this case); or in the MKS system, newtons per square meter, but rather (using the metric system as an example) kilograms divided by meters and then seconds squared. However, these are the same units (or dimensionality, in terms of where the units of mass, distance, and time are located) as there are for a force per unit area: Pressure is a force per unit area. The unit of force is the newton; for the unit of area we'd choose the square meter. A newton is also a kilogram accelerated at a meter per second, per second, or Kg-m/sec^2. Dividing that by one square meter in area gives Kg/m-sec^2, or the exact same thing. QED.

So...if you were surprised by how your arm suddenly got shoved back at 60 miles per hour (or however fast one of your parents happened to be driving at the time) when you rotated your palm ninety degrees from edge-on "flight" to broadside, imagine what it must be like for a space vehicle accelerating to orbital velocity. This "max q" is why the Space Shuttle usually throttled back to as low as 65% power below 35,000 feet (after which point it can then throttle back up). However, this equation is *not* the only thing that folks at the Johnson Space Center need to worry about! Above around 200 knots, other things start showing up. If you were to just plug some numbers in there though...at sea level, the density of air is roughly 1.225 Kg per cubic meter, or in the more clumsy English units, 0.002378 slug/ft3. (A slug is actually the amusingly proper term for a unit of mass in the British system, one slug "weighing" 32 pounds, and being the amount of mass that one pound of force would accelerate at one foot per second, per second.) The speed of sound at sea level is about 340.4 meters/second, or about 1116.9 feet per second. When you use these two quantities (regardless of the chosen units), you might determine that an object traveling at Mach 1 would feel a dynamic pressure roughly equivalent to 70% of normal everyday atmospheric sea level pressure (closest of course to what's given in choice B, though not exactly equal to it, as Peter Garrison implied in *FLYING* magazine not long ago). In the metric system, it would be about 70,972 Kg/m-sec^2. Since sea level pressure is about 101,325 newtons per square meter (or 2116.2 lb per square foot, or the more familiar 14.7 pounds per square inch), that's close. According to *those* calculations, dynamic pressure would about equal sea level atmospheric pressure at around Mach 1.2. Regardless of just how precise the answer really is, it gives you an appreciation for how much power it takes to push through air resistance at high speed. But if you've read this far...

There are additional considerations involved in such arcane calculations. A simple linear interpolation from a table in Van Nostrand's Scientific Encyclopedia yields a dynamic pressure at 662 knots of approximately 27 in Hg, or about 0.9 (additional) atmosphere. (A physics professor at the University of Colorado had informed Peter Garrison of FLYING magazine, when he had asked him on my behalf, that this dynamic pressure equals ambient sea level pressure at 1.195 times the speed of sound.) On the other hand, Barry Schiff has a friend who is also a physics professor (at the University of California), and *he* says that one must also consider the conservation of something known as enthalpy (the total value of energy of a system when it is at constant pressure). *His* answer for when dynamic pressure equals one (additional) atmosphere is: Mach 1.046...close enough. So again, choice B would be the closest one.

Answer to Question #32: Mariotte wrote an aerodynamics paper titled "Following the Movement of the Waters" (Traité du Mouvement des Eaux et des Autres Corps Fluides), published (posthumously) in Paris in 1686. His interests were quite varied: mathematics (navigation and the problem of determining longitude), optics (refraction and the "blind spot" in vision), hydrostatics (barometric changes, and the relation between gas pressure and volume), and the laws of motion and impacts of solid bodies.

Early researchers studied wind not with tunnels, but "whirling tables" (swinging arms, automating what any child does with a rock on a string). Langley used such a device in the late 1880s and early 1890s. The Wright Brothers *did* construct a wind tunnel in the fall of 1901, and data derived from their tests were vital to the successful of the 1903 Wright Flyer, but they were *not* the first to propose one. (For example, in 1896, Albert J. Wells built a wind tunnel at MIT, for his mechanical engineering thesis, and even earlier, in 1871, one was built by Frank H. Wenham. (At the first meeting of the Aeronautical Society of Great Britain, an engineer named Francis Herbert Wenham delivered a paper entitled "Aerial Locomotion," in which he drew the prescient conclusion that long, narrow wings would be most efficient. He had concluded this based upon measurements taken within his new invention, the wind tunnel, which he built in 1871 with John Browning.) However, the answer to the question, as it was worded, is D. (The answer to the question of who first *built* one would of course not be choice C, but choice A.)

Answer to Question #33: It's choice A. As an airship rises higher where the air is less dense and where the external pressure decreases, the helium within the ship expands and the cells swell out. At a point called "pressure height," when the cells become 100% full, any further ascent would result in increased pressure within the cells. If this were not relieved, it might eventually deform the surrounding structure and burst the cells. To guard against this, the cells are equipped with automatic valves set to open at a safe pressure and discharge their helium into the air. The Fujifilm web site had a pretty good "blimp glossary" if you're interested.

Answer to Question #34: D. The 727 plunged earthward at an average vertical speed of 46,000 ft (14,020m) per minute, with the rate of descent at moments reaching 76,000 ft (23,165 m) per minute. At that speed the aircraft broke the sound barrier and sonic booms were heard on the ground. It was the first time in history that a commercial airliner (other than the Concorde or the Soviet Tupolev Tu-144) broke the sound barrier. The airplane landed safely, and everybody on board survived. The captain immediately became a hero. He was given a medal by ALPA, and the news media honored him. But the praise did not last long. The flight crew was later accused by the NTSB of having purposely manipulated the flap / slat controls during cruise to eke out a little extra speed, knowing well that such a maneuver in cruise was against company procedure. The NTSB released the following accident report:

> "The safety Board determines that the probable cause of this accident was the isolation of the No 7 leading slat in the fully or partially extended position after an extension of the Nos 2, 3, 6 and 7 leading edge slats and the subsequent retraction of the Nos 2, 3 and 6 slats, and the captain's untimely flight control inputs to counter the roll resulting from the slat asymmetry. Contributing to the cause was a preexisting misalignment of the No 7 slat which, when combined with the cruise condition air loads, precluded retraction of that slat. After eliminating all probable individual or combined mechanical failures or malfunctions which could lead to slat extension, the Safety Board determined that the extension of the slats was the result of the flight crew's manipulation of the flap / slat controls. Contributing to the captain's untimely use of the flight controls was distraction due probably to his efforts to rectify the source of the control problem."

The NTSB blamed the accident on the crew. Within 24 hours, the crew, once praised as heroes, became the victims of an accusation they might have never committed. Till this day the NTSB maintains that the pilots were at fault. ALPA, on the other hand, still maintains their innocence. Urban legend has it that that captain's name was Hoot Gibson. Well, it wasn't. His last name *was* Gibson, but it was Harvey G. Gibson, not Robert L. "Hoot" Gibson, who later became a US astronaut (coincidentally, in that same year).

Chapter 5: Strange Weather

While it is unlikely that you'll ever have to factor Arctic temperatures into your altimeter corrections, or that you will ever want to know the exact formula for doing so, you'll be surprised how profoundly low temperatures affect how high you actually are (as opposed to how high you *thought* you were). And while you may not exactly have been waiting in breathless anticipation to learn that water vapor is lighter than air, or that the latent heat of fusion is partially responsible for the relationship between snowfall and atmospheric stability, there are useful insights behind some of the more obscure observations in the science of meteorology. It is these little known things which may be of some use to you later on.

Subject: Water, Water, Everywhere:

Question 1: Under what circumstances is water lighter than air?

A) Under conditions of extreme pressure, water will compress ever so slightly. Air however will compress until it is a liquid, and at a pressure of 610 atmospheres, it has a specific gravity greater than one.
B) When it's water vapor.
C) Inside a waterspout, the extreme low pressure actually causes water to exist in a quasi-gaseous state where, although technically still a liquid, its specific gravity becomes about one ten-thousandth that of liquid water on the surface--about ten times lighter than air!
D) Don't be ridiculous. Water is <u>never</u> lighter than air!

Subject: Ice Nine?

Question 2: Does ice always melt at zero degrees C, and does water always freeze at zero degrees C?

A) At normal atmospheric pressures (and up to and including 1000 atmospheres or so), yes.
B) The same as A, however water doesn't always freeze at zero C. At up to 70 Centigrade degrees below "freezing" however, and about 1000 atmospheres, water will stay liquid and in fact separate into two separate densities, one about 15% less dense than the other.
C) Although some weird stuff (like the above) happens at very high pressures, even here on normal old Earth, water can supercool if there are no condensation nucleii. Supercooled cloud droplets at -50 degrees C have in fact been observed, albeit rarely.
D) It depends on wind. If there is any advection (a fancy word for wind) greater than about 10 knots, supercooled water cannot exist.

Subject: Snowflakes

Question 3: They say no two snowflakes ever look alike. Aside from the obvious question of whether or not it's true (which it is), are there other types of snowflakes besides the familiar flat, frilly, dendritic six-sided "doily" shape?

A) There is only one basic shape for snowflakes. Their variety however, is infinite.
B) Snowflakes can come in three forms. They are the familiar hexagonal "flake", hairy "spheres" when condensation nucleii are over 50 microns and air temperatures are below 30 degrees C, and at about -4C, needles.
C) The number of shapes, as well as the variety within each, are both infinite.
D) Depending on temperature there are actually at least five types, although all are basically hexagonal in form. Between zero and -50F, snowflakes can be hexagonal tile shapes, needles, hollow prismatic tubes, tubes with hexagonal plates at either end, or the familiar frilly one with six (sometimes 12) arms.

Subject: Topsy-turvy

Question 4: True or false: It is possible for cloud bases to be above the cloud tops.

Subject: Brrrrrr!

Question 5: What is the origin of the term "wind chill"?

A) The origins are in basic atmospheric physics. There is a simple quantitative relationship between rate of air flow and potential heat exchange when water changes state across a fixed cross-sectional area. However, it was the Norwegian meteorologist Jakob Bjerknes who coined the term in 1935.
B) Although there was a formula used to calculate it, namely
T(wc) = wind chill temp = 91.4 - (0.474677 - 0.020425 * V + 0.303107 * sqrt(V)) * (91.4 - T)
where V is mph, and both T and T(wc) are in degrees F, there were several inherent errors, including the fact that the human body produces its own heat and that wind speeds are measured at weather stations 10 meters above the ground, where they can be much greater than they are at a height of five feet. (The old method estimated wind chill as colder than it really was, and a revision was implemented on Nov. 1, 2001.) The idea behind the "wind chill" figure was actually fairly recent. It is generally credited to the Austrailian Meteorology Bureau, and relatively recently, too: about 1955.
C) The term is credited to good old Ben Franklin. His "Poor Richards Almanack" in 1733 used the term (which reflects on how basically intuitive the concept actually is).

D) not that early! It is beleived that the English chemist and physicist, John Dalton, initiated the concept in about the year 1800.
E) To be precise, the term goes back to 1939. Antarctic explorer Paul A. Siple coined it his 1939 dissertation, "Adaptation of the Explorer to the Climate of Antarctica." (Siple had been one of Admiral Byrd's technical staff during his 1928-1929 explorations.) During the 1940s, both Siple and Charles F. Passel developed the wind chill index when they conducted experiments on the time needed to freeze 250 grams of water in a plastic cylinder that was exposed to the elements, hanging on a pole 33 feet high. They found that the time depended on how warm the water was, the outside temperature, and, of course, the wind speed.

Subject: Another World Of Cold

Question 6: What happens when it gets really cold, as in 60 degrees below zero?

A) Your breath freezes as you exhale, making a soft hissing noise.
B) Your eyelashes can freeze together
C) The slightest wind (5 kts) can induce frostbite on exposed skin in under half a minute.
D) all three

Subject: Another Angle On Fronts

Question 7: Approximately what are the actual slopes of cold and warm fronts?

A) Warm fronts slope up (in the direction of movement) at about a 30 degree angle. Cold fronts slope more, although down (in the direction of movement) and more steeply, at about a 45 degree angle.
B) A warm front slopes up only at a ratio of about 20 to one, while a cold front is a wedge of air about twice as steep, or 10 to one (again in the opposite direction).
C) Actually, fronts are really quite shallow. It's more like 1:40 for cold fronts, and 1:80 for warm fronts.
D) still too steep: cold fronts average roughly 1:100, warm fronts 1:200

Subject: High Lows and Low Highs

Question 8: Which unexpected coincidental relationship exists between Alaska and Hawaii regarding extremes of weather?

A) The highest temperature ever recorded in Alaska is not less than the highest temperature ever recorded in Hawaii.
B) The lowest temperature ever recorded in Hawaii is greater than the highest temperature on record for Alaska.
C) The highest wind speeds ever recorded in both states are identical.

D) Both states hold records for the highest barometer reading in the 20th century, and the barometric pressures were equal.

Subject: "When You're A Jet..."

Question 9: What is the minimum altitude for the jet stream?

A) 30,000 feet
B) by definition, 18,000 feet
C) 3000 feet
D) the earth's surface

Subject: I'll Huff, And I'll Puff...

Question 10: The highest microburst wind speed ever recorded occurred at what location?

A) Mount Washington, NH
B) Manhattan, KS
C) Houston, TX
D) Andrews Air Force Base, MD

Subject: God's Gatling Gun

Question 11: The growth of hail stones can continue as long as there are updrafts strong enough to keep them aloft. What is the largest hail stone ever recorded?

A) On March 10, 1939, at Hyderabad India, the heaviest hailstone ever recorded fell, and it weighed 7.5 pounds. (The diameter was not reported, however this would imply a diameter of approximately nine inches.)
B) on September 3, 1970, in Coffeyville, Kentucky, about 5.5 inches in diameter (slightly larger than a softball), weighing 1.67 pounds, and it was estimated to have struck the ground at 105 MPH
C) in Potter, Nebraska, on July 6, 1928: its diameter just under 5 and a half inches, and its weight was 1.51 pounds
D) On April 17, 1998, 4.5 inch diameter hail fell in Carnes, Mississippi, and there were unofficial reports of golf ball to softball sized hail in other locations.

Subject: Biblical Proportions

Question 12: About how much water falls as rain from an average rainstorm over one square mile?

A) 2,000 gallons
B) 20,000 gallons

C) 200,000 gallons
D) 2,000,000 gallons

Subject: A Mighty Day

Question 13: The greatest number of tornadoes ever to hit within a 24-hour period was

A) 25
B) 56
C) 107
D) 148

Subject: You Mean We're Not In Kansas Anymore?

Question 14: During the month of March, in which of these states are you most likely to see a tornado?

A) Oregon
B) Alabama
C) Ohio
D) Kansas

Subject: Okay, You Made This Word Up, Right?

Question 15: A "prester" is another name for

A) a dust devil
B) a waterspout
C) a tsunami
D) a chinook wind

Subject: Land of the Gentle Zephyr

Question 16: With the new Sport Pilot regulations, those folks who enjoy flying ultralights, powered parachutes, and any other aircraft that are less tolerant of gusts and sudden upsets, it might be useful to know where in the US one might find the lowest (or to put it another way) most gentle winds. Hence this question: Which city in the United States is the least windy?

A) Medford, Oregon
B) Jackson, Mississippi
C) Sacramento, California
D) Oak Ridge, Tennessee

Subject: The Windiest City

Question 17: Which US city has the highest average annual wind speed?

A) Amarillo, Texas
B) Dodge City, Kansas
C) Chicago, Illinois
D) Rochester, Minnesota
E) Boston, Mass

Subject: And I Suppose You Didn't Make This One Up Either, Huh?

Question 18: A "haboob" is

A) a Saudi sidewinder
B) a dust or sand storm
C) the aeronautical equivalent of a scupper hole for drainage of water from the edge of a deck, but in the case of aircraft it is from the lowest point in a fuselage
D) a garlic-flavored dip made from eggplant and tahini

Subject: Clocking Raindrops

Question 19: How fast does the average raindrop actually fall?

A) six knots
B) 10 knots
C) 16 knots
D) 32 knots

Subject: Fire and Rain (sorry, JT)

Question 20: True or false: Large forest fires can often occur under "cloudy skies".

Subject: Slightly Higher In Canada

Question 21: You're visiting Canada and flying over northern Manitoba in mid-winter. Your OAT (outside air temperature) reads minus 30 degrees Centigrade. About what in particular should you be worrying?

A) Your altimeter could read much higher than your actual altitude.
B) Your fuel could freeze.
C) The wings and propellers would get a better bite out of the denser air, but it requires a higher power setting.
D) A and B

Subject: More Strange Weather Words

Question 22: The word 'fulgurite' is used for which of the following:

A) any basic truism regarding weather, used in honor of Robert Fulghum's basic precepts on life
B) water which is dropped, usually from an airplane, in freezing air, and which freezes in curtain-shaped formations
C) another form of precipitation, only much more rare: rock... Fulgurites form when lava is explosively ejected and the molten spray cools before reaching the gound.
D) the instant formation of glass when lightning strikes a beach or desert floor

Subject: Who Takes the Cake?

Question 23: Tornadoes are the most severe weather phenomena. Which country can therefore be said to have the "worst weather in the world" because it has the most tornadoes?

A) China
B) the former Soviet Union
C) the United States
D) India

Subject: The Size Of Thor's Quiver

Question 24: How wide is a lightning bolt?

A) two inches
B) one foot
C) two feet
D) six feet

Subject: Thor's Attic

Question 25: To completely escape the possibility of a lightning strike in the vicinity of a thunderstorm, an aircraft would have to climb how high?

A) 10 miles: lightning strikes above 50,000 feet are virtually nonexistent
B) since lightning only strikes downward or horizontally, any altitude above the highest clouds within a radius of approximately 20 miles
C) FL 600
D) sixty miles

Subject: Catastrophic Visions Of Biblical Proportions?

Question 26: Can more than one tornadic event occur in the same general location at once?

A) Yes, several waterspouts can in fact form in close proximity to one another.
B) Yes, but never more than two at once, although it is not possible to have separations under two miles, due to the balance of pressure gradient and centripetal forces, and more importantly, the friction between adjacent funnels.
C) Yes, but no closer together than about three miles in very severe "supercell" storms.
D) No, not if you define "same location" as anywhere within five miles.
E) No. There has never been more than one tornado sighted within visual range of another.

Subject: Too Cold, Too Low

Question 27: Another trivia question mentions cold air altitude corrections for instrument approaches up where the air is really cold, like Canada, during the winter. However it wouldn't take arctic air to make you bust minimums. In the case where the surface temperature is just a bit below freezing, say minus 10 degrees C, how much of a cold temperature altitude correction should you add to all published altitudes, on any approach?

A) 20 feet if below 800 feet AGL, 40 feet between 800 and 1600 feet
B) three percent of your estimated AGL height below 1000 feet, six percent between 1000 and 2000 feet AGL
C) For an airport elevation below 2000 feet MSL, it is exactly ten percent of your height above the airport, up to 2000 feet AGL.
D) The bust would be technical, and legally valid, but negligible: five feet for a 200 foot DH; 15 feet for a 600-foot MDA; and 20 feet for an 800-foot MDA. Since your altimeter can legally be off by 75 feet anyway, this is indeed trivial!

Subject: The Driest Continent

Question 28: Which of the following is the driest continent on the planet?

A) Australia
B) Africa
C) Antarctica
D) The North Pole

Subject: Not Architectural, Neurological, Silvicultural, or Ceramic

Question 29: What aviation related item should come to mind when you see the terms "columns", "dendrites", "needles", and "plates"?

A) descriptions of different lightning shapes
B) types of mountain ranges
C) various possible snowflake shapes
D) cloud formations

Subject: How *Low* Can You *Go*?

Question 30: The lowest sea level barometric pressure (or altimeter setting) ever recorded in the United States was 26.35 inches. In which state was it measured?

A) Alaska
B) Florida
C) Oklahoma
D) Texas

Subject: Oh, Just Great...More Tornado Paranoia

Question 31: True or False: Tornadoes are always accompanied by visible funnel clouds.

Subject: The Perfect Storm

Question 32: The celebrated storm that took the lives of all aboard the *Andrea Gail* in the North Atlantic off the Canadian maritime provinces in late October of 1991 (and which was the subject of the movie "The Perfect Storm") was more than just an average example of the term "Nor'easter". Due to the cold front that was moving out into the North Atlantic off New England, a near-record intensity high-pressure system behind it with its associated gyre of clockwise motion and intensified winds from the northeast in that vicinity, *and* Hurricane Grace moving up the Atlantic coast, which pumped even *more* moisture into that frontal low, National Weather Service meteorologists have considered this event to be the worst of the Nor'easters. But just how did the term "Nor'easter" originate?

A) The term was coined by a NWS meteorologist in 1957 and signifies the *northeast movement* of any low pressure system having a pressure drop of more than 24 millibars in 24 hours. Contrary to popular belief, a nor'easter can exist by definition as far south as the Bahamas, and their associated winds may be from *any* direction, not just from the northeast.
B) Not only was the phrase "the perfect storm" coined by a National Weather Service meteorologist (in 1991), but in another one of those hokey quirks of fate where you might wonder which came first, the person's name, or his career choice, the meteorologist to whom we owe

this term was named Norris Easterman (but it was in 1969, not 1957, just before the Weather Bureau changed its name to the National Weather Service).

C) interesting, but almost total bilge water. These strong areas of low pressure, which while they may not equal hurricanes in intensity, often exceed them in duration, are so named simply because they can form in the North Atlantic off New England, and as such (being areas of low pressure with counter-clockwise flow) often drive winds ashore from a northeasterly direction as the storm approaches shore.

D) It isn't nearly so singular or distinctive. About the only thing that distinguishes a nor'easter is just that it's a plain old winter storm that happens to pick the northeast United States as the neighborhood in which to roost or run amok.

Subject: The Atmospheric Sponge

Question 33: On an average day (averaged over the entire year, that is) about how much water is in the atmosphere, just over the United States?

A) forty million gallons
B) forty billion gallons
C) forty trillion gallons
D) forty quadrillion gallons

Subject: Hot and Cold Running Water

Question 34: Just how cold and how hot can rain get?

A) from -40C up to 100C
B) from about -5C to about 40C
C) from 0C to 23C
D) from 5C to 30C

Subject: Raindrops Keep Falling On My Head...

Question 35: Galileo notwithstanding, with raindrops, the bigger they are, the faster they fall. (For objects of that size, the surface area to volume ratio dictates the extent to which it will overcome air resistance and viscous drag as it falls. Larger drops have a greater terminal velocity.) But when a raindrop falls faster than about 18 miles per hour (in still air) what will usually happen?

A) It will evaporate
B) This is the point where raindrops change shape from a relatively flattened (oblate) blob of water to the familiar teardrop shape.
C) It will break up into smaller raindrops.
D) Nothing happens. This is a trick question.

Subject: Let It Snow

Question 36: Fifteen minutes from your destination at 7,500 feet, and 500 feet below the clouds, you notice that snow is beginning to fall, and quite heavily. You're not instrument rated. You begin an immediate descent at a nominal 500 feet per minute. Although the speed that snowflakes fall varies, it averages out at roughly five mph. Assume that the snow is widespread and falls uniformly at that rate, your destination runway is at sea level, and you decide it's safe to make a straight-in arrival, to avoid possible instrument conditions. How much time, if any, will you have to land before snow begins falling on the runway and visibility could go below VFR minimums?

A) You won't make it. (And if it's heavy snow, you might as well be inside the clouds.) The snow will reach the runway five minutes ahead of you.
B) You'll have plenty of time, and you'd have over ten minutes to land, taxi in, and tie down. You could even make a normal pattern entry and relax.
C) You'd probably have about two minutes to spare.
D) You would make it, but the snowfall would be right above you, all the way down.

Subject: Hurricanes

Question 37: How many hurricanes strike the continental United States each year, on average?

A) 10
B) six
C) four
D) two

Subject: *More* About Hurricanes

Question 38: If a hurricane dissipates back into a tropical storm, but then regains strength and redevelops back into a hurricane, does it

A) get a new name
B) keep the old one
C) remain unnamed
D) This is a trick question; dissipated hurricanes never regain strength.

Subject: Raising Arizona?

Question 39: How many states in the entire US have had the honor of a summertime temperature record of 120 degrees Fahrenheit, or higher?

A) just one: Arizona
B) three: only Arizona, California, and New Mexico

C) 10
D) every single one

Subject: I'm Absoutely Parched!

Question 40: The driest place on earth, in terms of annual precipitation, is:

A) The Sahara Desert
B) Death Valley, CA
C) the Gobi Desert
D) Antofagasta, Chile

Subject: Damnation Alley, or Chicken Little?

Question 41**:** True or false: Global warming will inevitably cause more severe weather phenomena, and on a global scale.

Strange Weather: Answers

Answer to Question #1: The answer is choice B. When you think about the atomic numbers of hydrogen, oxygen, and nitrogen, it makes sense, after all!

Answer to Question #2: It's C, of course. Ice, at normal atmospheric pressure always melts when warmed to temperatures above zero degrees C. But water doesn't always freeze when cooled to 0C. Water in very small drops may take some tens of minutes to freeze when the temperature drops below -50 degrees C or so. Water droplets, even rain-sized drops, can exist in liquid form well below "freezing" if they do not have a freezing nucleus. The ice crystals "can't get started". After the water temperature goes below zero Celsius, the temperature decline is slow but significant. At some temperature below the "freezing point," the water finally does freeze and the temperature rises rapidly to zero Celsius. This means that heat is released from the freezing of water. There are around 80 calories of heat released for every gram of water which freezes. Just as heat is released in the condensation of water, freezing also causes heating (though less than one sixth as much, per gram). What this means though, is that to some extent, the presence of snowflakes implies that an additional boost of heat is transferred to the air in the form of updrafts (and downdrafts when it melts).

Answer to Question #3: It's choice D. And the reason most people DON'T know about the other snowflakes (aside from the fact that the other types aren't nearly as photogenic) is that those "grow best" at a range of temperatures (-12 to -16 degrees C) where the difference in vapor pressure over ice and liquid water is at its maximum. At 12C, *the most* energy is released when liquid water evaporates and condenses on ice crystals.

Answer to Question #4: True! Cloud bases are reported from airports as a height above ground level (AGL). Cloud tops on the other hand are reported as a height above mean sea level (MSL). For an airport like Furnace Creek, in Death Valley CA ("minus 210" on the sectional--assuming they ever have clouds, of course) if there were dense low clouds at 100 AGL that were just 100 feet thick, tops might be PIREP-ed at "sea level", ostensibly 100 feet below their own base! Incidentally there aren't many "below sea level" airports. Amsterdam's Schiophol is a few feet below sea level. The lowest of the low however is over by the Dead Sea. The elevation there at Metzada Airport (identifier LLMZ, and it's actually pronounced "msada") is 1,266 feet below sea level. Since ceilings and tops are reported over there the same way they're reported here, it would be theoretically possible (although there are no weather reporting facilities there).

Answer to Question #5: Choice E is the right answer. But it was revised as described in choice B, and for the reasons given.

Answer to Question #6: The answer is D.

Answer to Question #7: It's D. That makes sense when you think about the successively lower and lower cloud types that one might see over a period of two days, for example as a warm front moves eastward over a given point in the US. (So fronts really aren't nearly as narrow as one would think, when looking at them on Surface Analysis, Weather Depiction, or any other type of weather chart.)

Answer to Question #8: This one is choice A. Alaska's record high temperature was 100 degrees in Fort Yukon, on June 27 1915. (Its record low of -80F on January 23, 1971, is equally amazing.) Strangely enough, not only is the highest temperature ever recorded in Alaska *not less* than the highest temperature ever recorded in Hawaii, it's actually *equal* to it! Yes, the record high temperature for Hawaii is also 100F (at Pahala, on April 27, 1931). And Hawaii's lowest temperature was atop Mauna Kea (elevation 13,770 feet) on May 17, 1979, in case you're wondering.

Answer to Question #9: The right answer is choice D: The jet stream usually exists at altitudes between 30 to 40 thousand feet, but it can also reach as low as three to five thousand feet. (The official cut off" point at which such a river of upper level wind can be called a jet is if it is greater than 57 mph.) Jet streams approaching and even exceeding 250 mph are possible. The altitude of the jet stream depends on the vertical and horizontal temperature distribution in the atmosphere. During the winter when the temperature disparity between the equator and the poles is at its greatest, the jet stream generally reaches its maximum velocity. Low level jet streams are believed to be responsible for frequent high winds atop Mount Washington in New Hampshire, a site known for the "world's worst weather". (It also lies in the path of the principal storm tracks and air mass routes affecting the northeast United States.) The average winter wind speed at its summit is about 40 knots, and on April 12, 1934, a wind of 231mph was recorded there. All that said however, even lower streams have occurred; in these cases they are referred to as ground level jets.

Answer to Question #10: Answer: D. It occurred at 2:11 PM on August 1st, 1983, at Andrews Air Force Base---six minutes after president Reagan landed in Air Force One. Wind speeds reached about 120 knots.

Answer to Question #11: B. The event in choice A is urban legend and has not been substantiated.

Answer to Question #12: D.

Answer to Question #13: choice D: The worst outbreak of tornadoes in modern history occurred over Palm Sunday weekend on April 3rd and 4th, 1974, when 148 tornadoes hit 13 states, killed over 300 people, injured thousands more, and caused a half a billion dollars in property damage.

Answer to Question #14: It's B, in Alabama. Actually, different parts of the US have their own peak seasons for tornadoes. In the first three months of the year in fact, tornadoes are most common in the Southeast. This is where there are more confrontations between cold and warm air masses, and that is what helps to create the thunderstorms that produce tornadoes!

Answer to Question #15: B again. It's another name for a waterspout, which is just a tornado over water. Waterspouts incidentally move more slowly, due to their greater weight. (Water is heavy!) They also break up sooner, for the same reason (and the greater inertia of water which, when slung around in a circle, wants more to continue off in a straight line).

Answer to Question #16: It's D, Oak Ridge. The average wind in Oak Ridge, Tennessee is about four knots! Although this NOAA climate map doesn't reveal very low-level details such as city by city, as you can see this does agree on at least a regional level. Oak Ridge lies in a valley bounded by the Great Smoky Mountains to the southeast and the Cumberland Mountains to the northwest, which helps explain the shelter effect from prevailing westerly winds. That is one of the reasons that the United States' nuclear development was done at Oak Ridge; if there were an accident, the wind would not carry the radioactive particles as far as in other places.

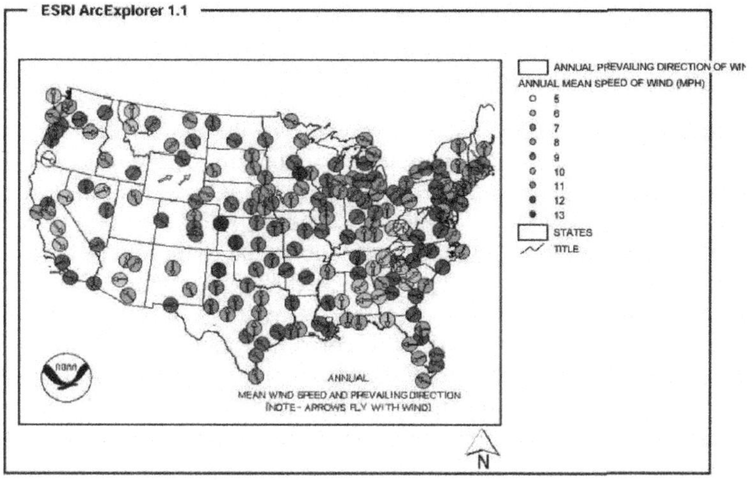

Image from ESRI ArcExplorer v 1.1
Reprinted with permission from NOAA, National Environmental Satellite, Data, and Information Service, National Climatic Data Center / Climate Atlas

Answer to Question #17: It's not Chicago, which although it claims the popular title of "The Windy City" actually has an overall average annual wind speed of about nine knots. In fact, getting out of Dodge is even easier (assuming the runway is favorably aligned with the prevailing winds) because the average winds in Dodge City Kansas are a bit over 12 knots. Yes, it's B.

Answer to Question #18: B (honest): From the Arabic habb, meaning 'to blow', it was originally a term for a penetrating sandstorm or duststorm in the deserts of Arabia, North Africa, and India, typically occurring late in the day during summer. However, they also occur in less exotic places like, say, Kansas, or Phoenix! They generally take a lens shape, appearing from a single point in their path as an advancing (and very intimidating) wall of sand or dust, generated from the surface outflow portion of a mature thunderstorm cell. They can move at speeds approaching 100 knots, and can last for several hours and rise to heights exceeding half a mile. In the US, they can occur from May through September, and most frequently in June. (Choice D is actually baba ganouj, in case you're a pita fan...)

Answer to Question #19: B. The vertical velocity of an average raindrop is about 10 knots (or 11 mph) without any wind. Larger drops can fall faster (about 16 knots), and very light mist can almost hover, falling at only about 1/10 knot.

Answer to Question #20: True. And they often do. Clouds need heat, and moisture, and a forest fire sure supplies both! These "forest fire" clouds even have a name: pyrocumulus. Moisture released from vegetation (and then driven up by the heat), combined with smoke particles, are what forms them.

Answer to Question #21: A. Flying into cold air is like flying into a low pressure area: you'll be lower than your altimeter says you are. The problem is that our altimeters cannot be corrected for temperature errors. On an instrument approach in actual IMC, this could be serious! Canadian pilots consult government-provided charts to determine how much altitude to add to those listed on approach procedures. (The U.S. Defense Mapping Agency has an altitude correction table for military pilots, and added only recently in the Aeronautical Information Manual, Section 7-2-3 has a temperature correction table.) At higher elevations and even colder temperatures, such corrections can approach 1000 feet! As to choice B, it would have to be below minus 55 C before you'd need to start worrying about choking your fuel lines with 100LL slurpee. (Most likely, it would start in the fuel lines leading from your tanks.) According to specifications on world aviation fuels from ExxonMobil, the freezing temperature for Grade 80, 100, and 100LL is either -58 or -60 degrees C (depending on issuing agency and specification). In winter, engines and airplanes climb better on *less* power (not more).

Answer to Question #22: D. Sometimes thought of as "fossilized lightning bolts", they often take on tube shapes (like this one). Actually, two separate phenomena can make natural glass: meteorites and lightning. Glass made from a meteorite's collision with the Earth's surface is called meteoritic glass or tektite. Fulgurites can also take on the form of a glassy crust whenever lightning strikes a bare rock surface, such as often occurs on mountain peaks. Although fulgurites are fragile, being glass, they last a long time, and are actually used as paleoenvironmental indicators. (For example, many have been found in the Sahara, where there is now little precipitation and lightning activity, confirming that very different conditions existed there in prehistoric times.)

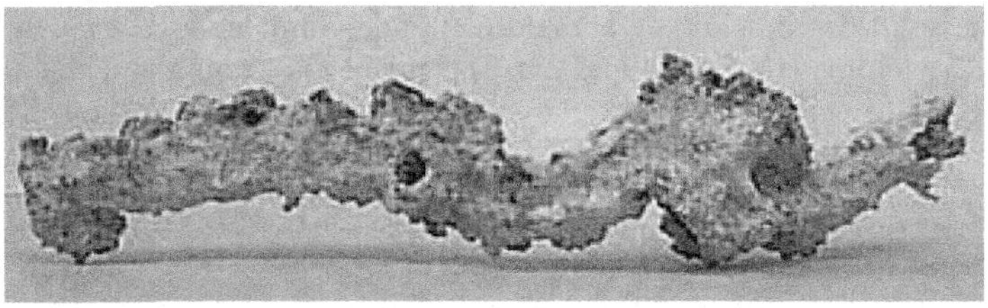

photograph of a "fossilized lightning bolt" or fulgurite

Answer to Question #23: It's C, the good old USA. As far as the occurrence of tornadoes goes, the US outnumbers China by 500-to-one, and the former USSR by 100-to-one.

Answer to Question #24: A. While the average lightning bolt is about two or three miles long, it is usually only somewhere between the thickness of a finger and about two inches in width. Although lightning can take other forms with much larger horizontal dimensions, lightning bolts are basically pretty skinny (but deadly, just the same).

Answer to Question #25: D. Yes, as ridiculously improbable as it sounds, 60 miles. We really *don't* know all there is *to* know about electricity in regards to weather. Although pilots have actually seen electrical discharges emanating upward from thunderstorm clouds for many years, it has only been fairly well documented within the last decade or so. Red and blue colored flashes of light from thunderstorms reaching altitudes as high as 60 miles have been recorded by NASA researchers aboard jet aircraft. Some extend up above the ozone layer into the ionosphere, where auroras occur. Our atmosphere may in fact be exchanging electric charges with the solar plasma in space, and storms just might be driven by energy from the larger environment through which Earth itself travels. (They may also be responsible for changes in the amounts of upper atmospheric ozone, for all we know at this point.) So far, two distinctly different flashes have been identified, named sprites and blue jets. Sprites are red flashes which last only a few thousandths of a second and extend from above storm clouds up to about 60 miles high, reaching the bottom of the ionosphere. Blue jets are flashes that appear in narrow beams or sprays resembling tracks of atomic particles or rays in a cloud chamber.

Answer to Question #26: A. In the case of waterspouts, adjacent multiple waterspouts *can* form along a cloud line, and violent tornadoes *can* develop as multiples, with one central large funnel, and smaller ones rotating (and revolving) around it. You can access a slew of these sorts of public domain images at "NOAA's Ark", the NOAA Photo Library (http://www.photolib.noaa.gov/historic/nws/nwind7.htm).

Multiple waterspouts off the Bahamas Islands, photo courtesy of the National Oceanic & Atmospheric Administration, Central Library (Photographer: Dr. Joseph Golden, NOAA)

Still don't think so? Here's a marine (not aviation) bulletin from the summer of 2003:

```
288
WHUS52 KTBW 041535
SMWTBW
GMZ830-853-041630-

BULLETIN - IMMEDIATE BROADCAST REQUESTED
SPECIAL MARINE WARNING FOR WATERSPOUTS
NATIONAL WEATHER SERVICE TAMPA BAY AREA - RUSKIN FL
1135 AM EDT MON AUG 4 2003

THE NATIONAL WEATHER SERVICE IN RUSKIN HAS ISSUED A

* SPECIAL MARINE WARNING FOR THE COASTAL WATERS FROM...
VENICE INLET TO JOHN'S PASS OUT TO 20 NM INCLUDING LOWER TAMPA
BAY. THIS INCLUDES THE COUNTIES OF SARASOTA...MANATEE...
HILLSBOROUGH AND PINELLAS.

* UNTIL 1230 PM EDT

* AT 1132 AM EDT...NATIONAL WEATHER SERVICE DOPPLER RADAR INDICATED
AN AREA OF THUNDERSTORMS...CONTAINING MULTIPLE WATERSPOUTS...FROM
7 MILES SOUTHWEST OF EGMONT KEY TO 7 MILES SOUTHWEST OF LONGBOAT
KEY...MOVING NORTHEAST AT 10 KNOTS.

AT 1130 AM EDT...A PILOT REPORTED MULITPLE WATERSPOUTS WITH THESE
STORMS.

A WATERSPOUT HAS BEEN SIGHTED! IF AT SEA...TAKE COVER BELOW DECK
UNTIL THE STORM HAS PASSED. OTHERS SHOULD REMAIN IN PORT UNTIL THE
STORM HAS PASSED.

&&
```

Answer to Question #27: The right one was C. (And from 3000 to 5000 feet AGL, the correction is just 10 feet less than 10 percent.) The reason behind this is that, again, altimeters aren't self-correcting for temperature. When it's cold, the air is more dense, and so the same air is "compressed" into a narrower layer. (Talk about density altitude!) The altimeter doesn't "know" that it's cold, as it only measures pressures, and it is calibrated for an "ISA" standard atmosphere, which is perfectly dry air at a sea level pressure of 29.92 inches Hg, at a much warmer sea level temperature of 15 degrees Centigrade, and a lapse rate of -1.98 C deg per 1000 feet. If you want to see the exact correction formula, it's this:

$$C = \frac{h(15 - t_0)}{273 + t_0 - .5k(e + h)},$$

where

C is the altitude correction value
h is the height above the facility
k is the standard temperature lapse rate (.00198° C/ft or .0065° C/m)
e is the facility elevation
t is the facility temperature
t_0 is the facility temperature adjusted to mean sea level, $t_0 = t + ke$

text and formula for altimeter altitude temperature corrections

So...radar altimeters, anyone?

Answer to Question #28: Answer: C. First of all, the North Pole isn't a continent; under that ice cap, there's just water, and no land. Secondly, even though most people imagine Australia to be a pretty dry and barren place--conjuring up visions of the Outback--it isn't. They have a large central and western area which *is* desert (about one third of the entire continent), surrounded by a ring with a semiarid climate (about another third), but outside of that, it can be subtropical and tropical and around Melbourne, quite humid. (Australia is however the driest *inhabited* continent.) Africa has vast areas in its northern third which are quite dry, like the Sahara Desert, but it also has a large central band where the climate is tropical. The driest continent is Antarctica (which *is* a continent, and technically, also a desert). Even though it's mostly covered by water (snow and ice) the frigid air can hold only very little moisture, and only a very few inches of it fall each year.

Answer to Question #29: C. These are some of the possible shapes that ice crystals take when snowflakes form. For any of you rock hounds out there, you might've thought they could have also been mineralogical. Certain minerals, due to their internal atomic structure, align themselves in a much more limited variety of patterns that are usually particular to that one mineral (although the basic rules that govern these assemblies are currently not understood in useful detail). With freezing water vapor however, there is a great variety in snow crystal formation (but the physics behind the morphology of snowflakes is also not too well understood, either). There are other additional terms to describe snowflake shapes, as well. There are "prisms" (hollow and solid), so-called "stars", and even "cups". Which one depends upon the temperature as well as the moisture content of the air at the time.

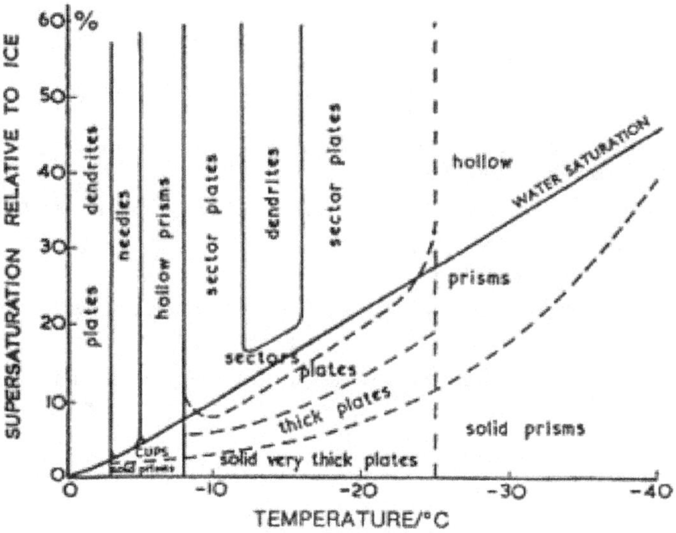

saturation vs. temperature graph for snowflake shapes

Answer to Question #30: Answer: B, Florida. Not that this is probably a big surprise, but it occurred during a hurricane. It was during the Florida Keys Hurricane of early September, 1935. (This was before they began naming them, which wasn't until after 1950.) It was also the lowest pressure ever recorded over land in North America. First though, a little background information...

As pilots, we know that the air gets less dense the higher you go, and also as the thermometer climbs higher. Even most non-pilots know that pressure decreases as we increase altitude, which makes jogging up in mile-high Denver Colorado such a challenge for an easterner. And as pilots (being the eager students of the atmosphere that we are), we also know that atmospheric disturbances can provide the equivalent of altitude, right here on earth.

The accepted means of determining how “thin” the air becomes is the use of a barometer to measure the pressure, which is an accurate measure of the weight per unit area of a column of air overlying the instrument. (Of course, we pilots must then correct for temperature, as well.) At sea level, on a normal day (if you’ll allow such unscientific imprecision), atmospheric pressure averages 14.7 pounds per square inch, which is sufficient to support a column of mercury 29.92 inches high in a glass tube. In metric units, this is expressed in terms of millibars. The standard sea-level pressure is 1013.25 millibars, which is the same as a force of 101,325 Newtons per square meter. Of course, on *cold* days, the density of the air increases, and for a given air mass, pressure rises. It’s also not a big surprise that higher barometric pressures are routinely observed in the Earth's deepest depressions. The “elevation” of the Dead Sea, along the boundary between Jordan and Israel, is 1286 feet below sea level. The highest pressure ever observed in this area was at Sedom, Israel (-1275 feet) on February 21, 1961: 31.94 inches (1081.8

mb). In comparison, the highest sea-level pressure in the United States was 107.86 kPa, or (since the number of millibars equals the number of kiloPascals times ten) 1078.6 mb, in Northway Alaska, on January 31, 1989. Canada's record for the highest sea-level pressure is a bit greater, at 107.95 kPa in Dawson YT on February 2, 1989. (So, okay, what's a Pascal? It's just one of those honorary units, each one signifying one Newton of force per square meter of area. What's a Newton? It's *another* honorary unit, this one being the force which will accelerate a kilogram of mass one meter per second, per second.) One might think that the South Pole, near the center of the Antarctic plateau, being that it routinely has such unimaginably cold temperatures, might take the cake for high density air. Nope…there's one problem: it's already high up. With a physical altitude of about 9,300 feet above sea level, the atmospheric pressure is already about 70% what it would be at sea level. The average pressure there is roughly 681.5 mb, or 20.125 inches of mercury. Ah, but there's another place where it's also cold, but not too high up: out in the Gulag of Mother Russia! The accepted world record maximum "sea-level equivalent" pressure was observed at Agata Lake (66 degrees 53 minutes North, 93 degrees 28 minutes East) in Siberia, at 1200 GMT on 31 December 1968. It was 32.01 inches Hg (1083.8 mb). The station elevation is 263 meters, and the temperature was -46 degrees C.

According to the Guinness Book of World Records, the lowest sea-level pressure ever recorded was 25.69 inches (87.00 kPa, or 870 mb) in the eye of Typhoon Tip. It was recorded 300 miles west of the island of Guam in the Pacific Ocean at latitude 16 degrees 44 minutes North, longitude 137 degrees 46 minutes East, on October 12, 1979. The U.S. record is 26.35 inches (89.23 kPA), produced by the 1935 Labor Day hurricane as it passed Long Key, Florida, at 10:00 p.m. on September 2, 1935. Other records were during Hurricane Camille, on August 17, 1969 (905 mb, or 26.72 inches Hg), during Hurricane Gilbert on September 14, 1988 (888 mb, or 26.22 inches), and Hurricane Andrew, on August 23, 1992, when the storm was still east of the Bahamas (922 mb, or 27.23 inches). In case you noticed that the low pressure for Gilbert was even less than that of the 1935 hurricane, and wondered why it wasn't listed as the all-time US record low pressure, that was because this occurred as it passed about 200 miles south of the western end of Cuba in the northwest Caribbean, and the worst of it missed the US altogether. (It made landfall in Cozumel, Mexico.) According to the Saffir/Simpson hurricane scale, the worst "Category 5" hurricanes (where winds are greater than 135 knots) have central pressures below 27.17 inches, or 920 mb. It is also possible that even lower pressures occur (however briefly) during the passage of a strong tornado, but so far, *no direct barometric pressure measurement* has ever recorded the lowest pressure at ground level in the center of a tornado. (A recent commercial venture involving "hardened in situ tornado pressure recorders" has recently begun operations, although meteorologists involved with tornado research remain skeptical. Due to the extreme winds, most measuring instruments are usually destroyed or badly damaged.) Pressures could be estimated with doppler radar, but the drawback is that this isn't at ground level. In addition, the very lowest pressure inside a tornado is at the top of the vortex, rather than near the ground.)

Answer to Question #31: False. Tornadoes can occur without funnel clouds, as "shown" in this image. Here, the dust cloud and cloud base above it were rotating, indicating a continuous cloud-to-ground vortex. The lack of a visible funnel can be related to several processes. One aspect is that before the funnel cloud touches down, it hasn't yet had a chance to ingest dirt and debris, and so it would not yet be as dark. Another very relevant factor is that the rotation and high tangential velocity of a funnel cloud are also accompanied by a significant drop in pressure, which often results in the air being cooled to its dew point, where moisture condenses. But if the air that feeds into the tornado is relatively dry, there will not be enough moisture to make the funnel cloud visible. In this case, it was likely that either the pressure drop and lift in the tornado vortex was too weak to cool and condense a visible funnel, or the air below the cloud base was just too dry (or both). All that said, generally, tornadoes without *any* funnel cloud are, thankfully, in the minority.

It is still most important to remember however that in the vortex of a tornado, the pressure is *relatively* uniform along its entire height, but the temperature profile is usually much steeper, being higher near the ground than near the top of the vortex. This is the *main reason* that the funnel clouds that are the most easily recognizable parts of a tornado usually seem to descend from the cloud base. However, in fact a vortex may *already* extend from inside the cloud, all the way down to the ground. When the air pressure inside the vortex drops, due to increased wind speed, it causes the ambient moisture in the air to condense. As the pressure inside the vortex drops, the moisture condenses, often beginning at the top of the vortex, where the temperature is lowest, and continuing downwards, which could give an observer the impression that the funnel cloud is descending. However, if the temperature is warmer near the *top* of the vortex, surprise! The funnel cloud suddenly appears and begins to rise to join the cloud base.

photograph of tornado forming "bottom up" without the initial warning of a funnel cloud descending from a cloud base (also occurring in dry air)

Answer to Question #32: It's C. The term indeed comes from the continuous and strong winds that blow from the northeast, and they can be among winter's worst storms. They actually begin as areas of low pressure in either the Gulf of Mexico or the Atlantic Ocean, which either moves up the East coast into New England and the Canadian maritime provinces, or out to sea. In New York or Boston, if the low moves offshore, and if the storm has enough moisture and cold air, they will see snow, heavy rain, or both, as well as high winds and beach-eating waves that crash onto the shore. These types of storms occur further south, though they occur most often in the northeast. But it's not where they happen so much as the direction from which the winds blow. The "almost total bilge water" comment was because there really *was* a NWS meteorologist (now retired) who originated that "perfect storm" thing in 1991; his name is Bob Case. (That other stuff about Norris Easterman, well, uh...)

Answer to Question #33: C. Including water droplets, ice, snow, and water vapor, averaged over the seasons, there are approximately 40,000,000,000,000 gallons of water poised over our heads. (That's forty trillion, using the definition of trillion as being ten to the twelfth power.) The quantity is more than that in the summer, when the air can hold more moisture, and less in the winter, but that's the average. Incidentally, on average also, about 10 percent of this water falls to the ground as precipitation.

Answer to Question #34: First, how cold can rain be? Well, let's get one thing straight: it ain't that much below freezing. Why? Because freezing rain is not the same thing as supercooled water. In situations involving freezing rain, precipitation that was originally snow melts, becoming rain aloft, and is above freezing as it falls. At some lower altitude is a (usually) shallow layer of air, which is below freezing. The raindrops fall through it so fast that their temperature doesn't drop below freezing, primarily due to water's high specific heat. The drops then hit a surface which is *below* freezing (like your airplane) where the water does turn into ice because it comes into contact (however violently) with that freezing surface. A significant percentage splashes off the ice at first, but a bit of it freezes on contact. Over time, that ice builds up and the water that splashed off finds a lower surface on which to freeze. There are freezing rain situations where the air temperature is about -5 C, but not much colder than that because if the temperature is very cold, the cold layer is likely to be quite deep. If so, the raindrops do freeze on the way down and instead become *sleet*.

Now, in the cloud itself, the water is supercooled. These are *very* small droplets and, in stratoform clouds, they are being lifted slowly (on the order of cm/sec). This is a perfect situation for supercooling. Nevertheless, at -40 C (also -40 F) supercooled water freezes spontaneously. So in that sense, the answer to how cold can liquid water be when you fly into it is: -40 degrees. But those supercooled droplets *never fall as rain*. Why? Well, they're about a million times smaller than a raindrop, and they just won't fall, *and* because in a cloud, the saturation vapor pressure with respect to ice is less than the saturation vapor pressure with respect to water—in order for droplets to form from pure water, the saturation must be around 120%, not just 100%—so the process (named the Bergeron process, incidentally) causes snowflakes to grow via sublimation at the expense of the supercooled water droplets, and it is the snowflakes which fall out of the cloud. Of course, if you fly through the super-cooled droplets, they will freeze onto the plane. So the difference between supercooled cloud droplets and freezing rain is academic. As far as pilots are concerned, the effect is the same: you get iced.

How warm can rain get? In theory, it could be almost 100 C! Such a theoretical case would be a bit extreme, but it's conceivable. If there were a situation such as the one described in another trivia item, in which pyrocumulus or "forest fire" clouds (formed by moisture released from vegetation during a forest fire and driven up by the heat) were mentioned, and they managed to release rain which fell through the superheated air, the raindrops could become quite hot. However, in all other more realistic scenarios, it's much less than that. In India, in April, it gets well into the 100-110 F range before the monsoon breaks, and after the rains arrive it cools down...all the way into the 90's! (It *is* possible in the tropics, where evaporative cooling is minimized, that the rainfall could be that warm.) But this has actually not been considered important enough to record accurately. The best answer is choice B.

Answer to Question #35: It's C. Raindrops are formed when the innumerable numbers of tiny droplets within clouds grow in size as moisture from the surrounding air condenses on them, or as they coalesce with other droplets during their descent. Raindrops vary in size from about 0.02 in. (0.5 mm) to about 0.2 inches (about 5.5 mm). Rarely, drops as large as 0.33 in. (8 mm) have been reported during thunderstorms, although this is not as well documented. (The appearance of such large raindrops is always associated with convective activity, turbulence, and strong updrafts, which would allow raindrops to continue to grow.) From the time they leave their parent cloud, evaporation occurs, and if the cloud is high, the air is dry, and the raindrops are small enough, they may evaporate completely, before reaching the ground. When that happens, it's called virga. Generally though, the "average" raindrop (which falls at about 11 mph) holds together until its velocity reaches about 18 mph (barring turbulence, the relative motion of updrafts or downdrafts, etc.) After that, it breaks up into smaller raindrops. They actually do change shape, except that up to about 2 mm in size, raindrops are generally spherical, after which they become more oblate and their bottoms flatten out, until they come to resemble the top half of a hamburger bun in shape. (In addition to that, they oscillate in shape somewhat as they fall.) Once they reach about a fifth of an inch (about 5.5 mm) in size, raindrops become unstable and they break up. So, the next time you're caught in a rainstorm and it feels like a fusillade of bullets, well...it really isn't that bad.

Answer to Question #36: Most snowflakes fall rather slowly, due to their relatively large surface area and low weight. Their average speed of about five miles per hour would about equal a 440 fpm descent. Descending from 7500 feet at 500 fpm takes 15 minutes, which would put you over the numbers just at the conclusion of your descent (if no other traffic intervened). During those 15 minutes, you would gain about a 900 foot lead on the snowfall, which would give you about two minutes to land and clear the runway. The answer is C. That's cutting it a bit close, but it could work. It would probably be smarter to descend more quickly, land at a closer airport, or both.

Answer to Question #37: Every year on average, there are 10 tropical storms that begin their life in the Atlantic basin. Out of those 10, six will become hurricanes. However, of those six, only about two (choice D) strike the U.S. mainland in any given year.

Answer to Question #38: First of all, the naming starts as soon as a tropical depression reaches tropical storm status. So reverting from hurricane strength back to a tropical storm doesn't change anything. Before anything ever gets to tropical storm status though, there are earlier and less sinister formative stages of development, the first of which is just simply a low pressure system. When there is a low pressure system that is non-frontal in origin, is fairly large (so called "synoptic scale"), it forms over tropical or sub-tropical waters and has organized convection (i.e., thunderstorm activity) as well as definite cyclonic surface wind circulation, the generic term used to describe it is the "tropical cyclone".

The answer is choice B, by the way.

Now you're probably wondering: what does he mean by cyclonic? Well, that just refers to rotation in the same direction as that of the Earth, as seen from above either pole. For the Northern Hemisphere, cyclonic rotation is counter-clockwise, and for the Southern Hemisphere, cyclonic rotation is clockwise. So these terms cyclonic and anti-cyclonic are relative; which way they rotate is related to the hemisphere in which they occur.

But back to that strength thing again: just when does a cyclone become a tropical storm, and when does a tropical storm become a hurricane? Tropical cyclones with maximum sustained surface winds of less than 34 knots (about 39 mph) are called "tropical depressions". Once a tropical cyclone reaches winds of this velocity however, it's then called a "tropical storm" and is assigned a name. If winds reach 64 knots (or 74 mph), then they start referring to it as a "hurricane". Note that this applies in the North Atlantic Ocean, the Northeast Pacific Ocean east of the dateline, and the South Pacific Ocean east of 160 degrees East longitude. For those of you in (or traveling to) the Northwest Pacific Ocean west of the dateline, the term is "typhoon"; in the Southwest Pacific Ocean west of 160E or Southeast Indian Ocean east of 90E, it's called a "severe tropical cyclone"; in the North Indian Ocean, the term is "severe cyclonic storm"; and in the Southwest Indian Ocean, it's a "tropical cyclone". (Clear as mud, right?)

And just what "maximum sustained surface winds" means depends on who is doing the measuring; World Meteorology Organization guidelines say that a 10 minute average defines "sustained", and most countries use this. But the National Hurricane Center and the Joint Typhoon Warning Center (JTWC) use a one minute period to compute sustained winds. This difference introduces some complications in comparing statistics from one ocean basin to another (because using a smaller time interval probably increases the number of occurrences).

By the way, names are reserved for tropical storms years in advance. If you'd like to see if your name was on the list of names recently used, have a look:

Atlantic Names

2002	2003	2004	2005	2006	2007
Arthur	Ana	Alex	Arlene	Alberto	Andrea
Bertha	Bill	Bonnie	Bret	Beryl	Barry
Cristobal	Claudette	Charley	Cindy	Chris	Chantal
Dolly	Danny	Danielle	Dennis	Debby	Dean
Edouard	Erika	Earl	Emily	Ernesto	Erin
Fay	Fabian	Frances	Franklin	Florence	Felix
Gustav	Grace	Gaston	Gert	Gordon	Gabrielle
Hanna	Henri	Hermine	Harvey	Helene	Humberto
Isidore	Isabel	Ivan	Irene	Isaac	Ingrid
Josephine	Juan	Jeanne	Jose	Joyce	Jerry
Kyle	Kate	Karl	Katrina	Kirk	Karen
Lili	Larry	Lisa	Lee	Leslie	Lorenzo
Marco	Mindy	Matthew	Maria	Michael	Melissa
Nana	Nicholas	Nicole	Nate	Nadine	Noel
Omar	Odette	Otto	Ophelia	Oscar	Olga
Paloma	Peter	Paula	Philippe	Patty	Pablo
Rene	Rose	Richard	Rita	Rafael	Rebekah
Sally	Sam	Shary	Stan	Sandy	Sebastien
Teddy	Teresa	Tomas	Tammy	Tony	Tanya
Vicky	Victor	Virginie	Vince	Valerie	Van
Wilfred	Wanda	Walter	Wilma	William	Wendy

Eastern North Pacific Names

2002	2003	2004	2005	2006	2007
Alma	Andres	Agatha	Adrian	Aletta	Alvin
Boris	Blanca	Blas	Beatriz	Bud	Barbara
Cristina	Carlos	Celia	Calvin	Carlotta	Cosme
Douglas	Dolores	Darby	Dora	Daniel	Dalila
Elida	Enrique	Estelle	Eugene	Emilia	Erick
Fausto	Felicia	Frank	Fernanda	Fabio	Flossie
Genevieve	Guillermo	Georgette	Greg	Gilma	Gil
Hernan	Hilda	Howard	Hilary	Hector	Henriette
Iselle	Ignacio	Isis	Irwin	Ileana	Ivo
Julio	Jimena	Javier	Jova	John	Juliette
Kenna	Kevin	Kay	Kenneth	Kristy	Kiko
Lowell	Linda	Lester	Lidia	Lane	Lorena
Marie	Marty	Madeline	Max	Miriam	Manuel
Norbert	Nora	Newton	Norma	Norman	Narda
Odile	Olaf	Orlene	Otis	Olivia	Octave
Polo	Patricia	Paine	Pilar	Paul	Priscilla
Rachel	Rick	Roslyn	Ramon	Rosa	Raymond
Simon	Sandra	Seymour	Selma	Sergio	Sonia
Trudy	Terry	Tina	Todd	Tara	Tico
Vance	Vivian	Virgil	Veronica	Vicente	Velma
Winnie	Waldo	Winifred	Wiley	Willa	Wallis
Xavier	Xina	Xavier	Xina	Xavier	Xina
Yolanda	York	Yolanda	York	Yolanda	York
Zeke	Zelda	Zeke	Zelda	Zeke	Zelda

tables of Atlantic and eastern North Pacific tropical storm names determined by a committee of the World Meteorological Organization (compiled and created by the author)

For hundreds of years, many hurricanes (at least in the West Indies) were named based on whatever saint's day a particular hurricane occurred. Clement Wragge, an Australian meteorologist, began giving women's names to tropical storms in the late 1800s. Starting in 1953, Atlantic tropical storms were named from lists originated by the National Hurricane Center. (They're now maintained and updated by an international committee of the World Meteorological Organization.) The practice of naming hurricanes after women only ended in 1978 when men's *and* women's names were included in the Eastern North Pacific storm lists (and in 1979, lists for the Atlantic and Gulf of Mexico as well). Six lists are now used in rotation, being "recycled" every six years. The one exception is when a storm is *so* deadly or costly that the future use of its name on a different storm would be insensitive to the loss of property and life, in which case the WMO committee removes it and chooses another name. (For example, on the 2004 list, Gaston has replaced Georges, and Matthew has replaced Mitch.)

Answer to Question #39: Answer: It's not global warming, but it's been hotter than most people realize. Fully 10 states have had the dubious honor. In alphabetical order, they are Arizona, Arkansas, California, Kansas, Nevada, New Mexico, North Dakota (yes, really), Oklahoma, South Dakota, and Texas. So the answer is: C. In fact, it might surprise you to learn that absolutely every state in the Union *has* had record temperatures of 100 degrees or more (even chilly Alaska and balmy Hawaii). Here is one compilation (although it may not be absolutely current)...see for yourself:

State	Temp	Date	Location
Alabama	112	Sept. 5, 1925	Centerville
Alaska	100	June 27, 1915	Ft. Yukon
Arizona	128	June 29, 1994	Lake Havasu
Arkansas	120	Aug. 10, 1936	Ozark
California	134	July 10, 1913	Death Valley
Colorado	118	July 11, 1888	Bennett
Connecticut	106	July 15, 1995	Danbury
Delaware	110	July 21, 1930	Millsboro
Florida	109	June 29, 1931	Monticello
Georgia	112	July 24, 1952	Louisville
Hawaii	100	April 27, 1931	Pahala
Idaho	118	July 28, 1934	Orofino
Illinois	117	July 14, 1954	E. St Louis
Indiana	116	July 14, 1936	Collegeville
Iowa	118	July 20, 1934	Keokuk
Kansas	121	July 24, 1936	Alton
Kentucky	114	July 28, 1930	Greensburg
Louisiana	114	Aug. 10, 1936	Plain Dealing
Maine	105	July 10, 1911	North Bridgton
Maryland	109	July 10, 1936	Cumberland and Frederick
Massachusetts	107	Aug. 2, 1975	New Bedford and Chester
Michigan	112	July 13, 1936	Mio
Minnesota	114	July 6, 1936	Moorhead
Mississippi	115	July 29, 1930	Holly Springs
Missouri	118	July 14, 1954	Warsaw and Union
Montana	117	July 5, 1937	Medicine Lake
Nebraska	118	July 24, 1936	Minden
Nevada	125	June 29, 1994	Laughlin
New Hampshire	106	July 4, 1911	Nashua
New Jersey	110	July 10, 1936	Runyon
New Mexico	122	June 27, 1994	Lakewood
New York	108	July 22, 1926	Troy
North Carolina	110	Aug. 21, 1983	Fayetteville
North Dakota	121	July 6, 1936	Steele
Ohio	113	July 21, 1934	Gallipolis
Oklahoma	120	June 27, 1994	Tipton
Oregon	119	Aug. 10, 1898	Pendleton
Pennsylvania	111	July 10, 1936	Phoenixville
Rhode Island	104	Aug. 2, 1975	Providence
South Carolina	111	June 28, 1954	Camden
South Dakota	120	July 5, 1936	Gannvalley
Tennessee	113	Aug. 9, 1930	Perryville
Texas	120	Aug. 12, 1936	Seymour
Utah	117	July 5, 1985	Saint George
Vermont	105	July 4, 1911	Vernon
Virginia	110	July 15, 1954	Balcony Falls
Washington	118	Aug. 5, 1961	Ice Harbor Dam
West Virginia	112	July 10, 1936	Martinsburg
Wisconsin	114	July 13, 1936	Wisconsin Dells
Wyoming	114	July 12, 1900	Basin

table of US states and maximum summertime temperature records
in degrees Fahrenheit, created by the author

Answer to Question #40: The Sahara Desert is pretty big, covering about 10 percent of the African continent. It's pretty dry, too; in its driest eastern parts, the average rainfall is only about five millimeters, or around a fifth of an inch per year. (Keep in mind this is an average figure; there might be no rain at all for many years, followed by a single intense thunderstorm.) And Death Valley, with its forbidding name, isn't too hospitable either, being famous for its dry climate and very hot summer temperatures. But compared to the Sahara, it's a tropical paradise; it's average annual rainfall is about 1.85 inches! The Gobi Desert of southern Mongolia isn't known for being terribly hot (in fact it can be quite cold) but it also has little rainfall, mostly because the Himalayas prevent moisture from moving into the region. It averages under three inches of rain per year. (To be called a desert generally speaking, a region must get under 10 inches of rain per year.) But the winner as far as being the one place in the world most devoid of precipitation is the dry Atacama desert region of Antofagasta (pronounced "ahn-toh-fa-GAH-stah", with the primary accent on its fourth syllable), in the generally dry northern region of Chile. Home to about a quarter of a million people, and despite the fact that it is also a port on the Pacific Ocean, the average annual rainfall there is almost nonexistent: less than one tenth of one millimeter. Several years can easily go by with absolutely no rain at all. (So, yes, it's choice D.) This part of the eastern Pacific *is* indeed quite parched, but they've come up with some clever means of obtaining fresh water. Along the arid coast of Chile as well as southern Peru, clouds settling on the Andean slopes produce what is known locally as camanchacas (thick fog). Such heavy mists which touch the land surface are actually "milked" to obtain water, by means of plastic sheeting hung on vertical posts, from which condensation then drips into conduits and is then collected. Such fog harvesting has actually been used to provide water for small rural communities in arid and semi-arid regions in many other parts of the world as well, such as Namibia, Mexico, the Canary Islands, Australia, and even California. Also, though this part of the world doesn't get much rain, at least it's not beastly hot. This is mostly because atmospheric temperatures are moderated by the counter-clockwise gyre of colder water coming from areas to the south; the offshore presence of the Humboldt or Peru current keeps the average temperatures at Antofagasta between about 65° to 75° F in January, and from 50° to 65° F in July.

photograph of "fog harvesting" sheets for extracting water from the camanchacas (thick fog) of Chile

Answer to Question #41: One argument goes that evidence supporting this has already been established by the World Meteorological Organization, as well as leading climatologists, worldwide. Some would assert that it is an undeniable fact that worldwide weather has indeed gone haywire, and that this is directly related to the planet's warming temperatures. Many people will insist that their own regions have had nothing but extreme, record-breaking, never-before-seen weather for the past several years. All over the world, from southern Argentina to Japan to sub-Saharan Africa, it seems that our climate is changing, right before our eyes.

Some further insist that the cause is our use of fossil fuels, and that only a rapid switch to clean energy, made possible by the involvement of enough concerned citizens, will save us from the more severe impacts that scientists believe are fast approaching. So, what's the deal? Are 120 degree summers in the offing for Pennsylvania? Will our grandchildren see F8 near-supersonic tornadoes in the midwest? Will Atlanta be a coastal city by 2100 AD?

True, there is a problem. Most climatologists do agree that we are experiencing global warming. What they *don't* agree on is the severity and the consequences of this warming. Record heat did occur in France (and all of western Europe) during the summer of 2003, and we in the US did have a lot of tornadoes during one recent year in May. However, records get broken every year. Extreme weather is actually "normal" in that it occurs more frequently than you'd expect. That's a problem: the general public expects weather to behave nicely, and it gets anxious when it doesn't. Statisticians expect data to have normal distributions and thus expect departures from the mean of two standard deviations or more only about 5% of the time. But the catch is that weather is *not normally distributed* such that extremes of two, three, and four standard deviations occur more frequently than 5%! This gives the public the impression the weather is going bananas when it is *just behaving properly for a data set with a non-normal distribution.*

This also obscures the real fact that we *are* changing the way the atmosphere behaves. Meteorologists are still not sure which extreme events would have occurred anyway in the absence of global warming. Furthermore, global warming leads to cold events in some places. Some conveniently forget to mention the cold, snowy winter in the eastern United States...which, incidentally, was also related to global warming through a strengthened jet stream. But beware of the Lorax! Screamers with agendas rarely want people to examine all the data in a rational way to come to scientific conclusions. As best as we can know at this point, the swamis say: false.

Chapter 6: Aerospace

Here we find orbital trajectories, flying in the Martian atmosphere, SR-71 Blackbirds, the well-known weaknesses of the Space Shuttle tiles, and the sorts of things that might be day-to-day concerns only for the astronaut hopefuls among us. Nonetheless, a few glances into the non-intuitive realm of the Very Fast can't help but be interesting for the rest of us slow pokes.

Subject: Speed Limits

Question 1: The SR-71 was typically flown at Mach 3.0 or 3.15, although in reality the engines were quite capable of much greater speeds. Why wasn't it flown any faster than this?

A) the need for refueling became too prohibitively frequent
B) the speed restriction in the FARs for flight above FL600
C) It would have melted.
D) While it wouldn't enter orbit, it would enter a parabolic arc high enough where control surfaces are no longer effective, and attitude could not longer remain controlled.

Subject: Private Pilot Question From the Twenty-Second Century

Question 2. Two pairs of aircraft are in formation flight. One pair is 10,000 feet, and the other pair is quite a bit higher; at one million feet (in other words, outside the earth's atmosphere about 190 miles up, in orbit). In each case, one aircraft is in front of the other. If the second aircraft in each formation wants to overtake the first, what must it do, in each case?

A) Both trailing aircraft need to add power (whether that means manifold pressure, jet power, or rocket power).
B) The aircraft in trail at 10,000 feet needs to speed up, just slightly. The aircraft in trail at "FL 10,000" actually needs to slow down in order to speed up.
C) Huh?
D) Formation flight in orbit is not allowed unless all vehicles involved have three-axis orbital maneuvering systems (FAR 91.1415).

Subject: Follow the Sun

Question 3: If you took off due West at a given time, and tried to follow the sun so that you "stayed with it"---whether that was noon, sunset, whatever--- below what latitude would you no longer be able to do so without exceeding the speed of sound?

A) Unless you could afford to charter a supersonic jet, or are one of the select few who owns a war surplus fighter that can exceed the speed of sound (or your commanding officer were to let you get real creative next time you took up one of Uncle Sam's F-16s on a ferry flight), it wouldn't be possible. The nadir point of the sun crosses longitude lines at about Mach 1.37.
B) Nice try, Maverick Weenie. I said "at what latitude", so the question does not assume that you're trying to keep the sun always overhead. Assuming you were to feel free to hug the 49th parallel, if you wanted to, and you just wanted to keep the sun somewhere above the horizon, you'd have to go to about 35 degrees of latitude, where the tangential rotation speed of the earth goes below 762 miles an hour.
C) Closer, but not quite right. The approximate tangential rotation speed of the earth is its equatorial radius times the cosine of your latitude, in turn multiplied by the earth's angular velocity. You would have to be at about the latitude of Buffalo, NY (actually 42.9 degrees North) in order for that to be below the speed of sound. And of course, if it were winter, and you wanted some reasonable elevation angle to keep the sun in your sights, you might want to try it at 42.9 degrees South...
D) Still not quite right. The speed of sound also varies with air temperature. At sea level, at standard temperature and pressure, that's about 762 mph, yes. But if you're flying something that fast, you might be up at 36,000 feet, above which point the air temperature stays a relatively constant -56 deg C, and where the speed of sound drops to about 660 mph (or about 573 knots). Limited at that speed, you would need to go even further North...a latitude just a bit above Moose Jaw, Saskatchewan, would do nicely: 50.6 degrees N.

Subject: Glass Jaw

Question 4: The Space Shuttles had tiles that can resist the terrific heat that is generated during re-entry through the earth's atmosphere, but what couldn't they stand up to? (More than one may apply.)

A) cold: They contracted and de-bonded at temperatures below -15 degrees C.
B) impact: They were very sensitive to touch and can actually crumble if subjected to any frictional forces.
C) rain: Unlike the Wicked Witch of the West, while they wouldn't shrivel up if someone tossed a bucket of water on them, the impact of rain drops in transit atop the Shuttle Carrier Aircraft traveling at between 300 and 400 knots would also cause structural damage to the tiles.
D) night: Darkness damages the tiles.

Subject: Looking Sharp, Moving Fast

Question 5: Hypersonic vehicles do not generally have sharp-edged wings or noses. Why not?

A) Shock waves above Mach 5 are parabolic, not angular, and so they don't need to be pointy. Blunter structures are stronger, not to mention cheaper to make.
B) Sharper and "pointier" objects have proportionally greater surface area to volume ratios. Therefore, they have more area to heat up, and less volume with which to absorb that heat. Even inconel, with a 1370 deg C melting point, cannot hold up to hypersonic flight when fashioned into something sharp, skinny, and pointy. That's why the front ends of the Space Shuttles look about as sexy and suave as a sea manatee.
C) They *can* be sharp and spiky, actually. That surface area to volume stuff is fuzzy thinking. The more surface area something has per unit volume, the easier it is for it to radiatively dissipate heat.
D) Surface insulation tiles, like the curvy brick jacket of silica fiber tiles sported by the Shuttle, can't take load deformation, which is sure as heck what they'd get if they stuck out into a Mach 5 breeze.

Subject: Three Greens While the World Watches

Question 6: The Space Shuttle typically didn't extend its landing gear until

A) it is about 300 feet above the runway
B) it has descended to within approximately 2000 feet above the runway
C) it is established on a seven-mile final
D) just below Mach 0.9 at 10,000 feet

Subject: Flying On Mars

Question 7: Off in the far azure infinity of man's spacefaring destiny, if he has one, aircraft may someday soar through the Martian skies. Due to the thin air however, they won't be Skyhawks. They will need a high aspect ratio (like 30-to-1) and "flying wing" proportions more like Helios, NASA's high-atmospheric solar vehicle that already nibbled at nearly 100,000 feet (which is just about what the Martian atmosphere is like at its surface, at least in terms of density). With a weight about equal to a Cessna 152, but with a nearly 250-foot wingspan, the thin air allowed Helios somewhere around a 140 knot ground speed at altitude (though its takeoff and landing speeds were more akin to those of a bicycle). Just for fun though, if someone could pull some kind of a trillionare "jackass" stunt and fly a rocket-powered 172 on Mars--it would *need* to be rocket powered, since any air-breathing engine would probably suffocate--how fast would it need to go?

A) Due to the fact that the gravity on Mars is only about one-third that of earth, takeoff and landing speeds would only be about twice what they are here on earth.
B) Indicated airspeeds would be exactly the same, but you'd need titanium tires (and a *very* smooth runway), because your takeoff speed would be somewhere around 500 knots
C) supersonic
D) impossibly fast: somewhere between 2300 and 2900 knots, depending on the Martian season

Subject: From the "Seeing Is Believing" Department...

Question 8: What is happening to the airplane in this picture?

U.S. Navy photo by Ensign John Gay

A) It is entering a well-defined warm front.
B) This picture was taken at the instant the pilot applied afterburners.
C) It was taken just as the airplane exceeded Mach 1.
D) This is what happens when someone throws a water balloon at a jet doing 500 knots.
E) It's nothing special actually. This jet is simply flying through a (very) small cloud.

Aerospace:
Answers

Answer to Question #1: The right answer is: C; temperature. It could be flown faster, but not for long! Although above 99% of the earth's atmosphere, frictional heating is still so great that before too long, critical surfaces of the mostly titanium skin---especially around the engines---would begin to soften and start coming apart. (This information was from a lecture at the College Park aviation museum by a former SR pilot, Tom Allison, as well as correspondences with Marta Bohn-Meyer.)

Answer to Question #2: Choice B is correct. Figuring out what to do in the case of the two aircraft at 10,000 feet isn't rocket science but for the two waltzing around in orbit, well, um, it kind of is. Outside the atmosphere, the two "aircraft" are really following ballistic trajectories, either in some sub-orbital parabola or in elliptical orbit. As such, Kepler's second law of conservation of angular momentum applies. The radii of each vehicle's orbit must sweep out equal areas in equal times. What this means is that, if the second craft were to add power at its six o'clock, it would move to a new (or higher) apogee (half an orbit later), and actually be at a greater altitude. Hence, it's ground track would then actually slow down as the earth rotated relatively faster beneath it. So unless you do a "re-or" first, you would actually move above and behind the vehicle in front of you (being furthest away about 45 minutes later). Nope, in orbit, the fact is that you must "fire in the direction that you want to go". The aircraft in trail must turn around, fire just a tinch in its direction of motion, after which it will attain a lower orbit (or "perigee") half an orbit later, and be travelling just a bit faster. Communication satellites in geosynchronous orbit which are periodically adjusted in longitude to stay "in the FCC box" don't need to reorient themselves first, as they have both East and West-facing thrusters, but they do just that very counterintuitive thing, in fact.

Answer to Question #3: The answer of course is D.

Answer to Question #4: B, and C. NASA used two specially modified Boeing 747s to ferry Space Shuttle orbiters back to the Kennedy Space Center from remote landing sites. These two billion dollar aircraft could withstand the searing heat of re-entry over and over again, but their tiles were sensitive to touch and debris--and that included being pelted by rain drops. Once mated to the carrier aircraft, they more than doubled the normal 300,000+ pound gross weight of the 747, double the fuel burn, and cut its speed by a third. And instead of a 5000 mile range, they typically flew 1000 mile legs. They couldn't fly through rain, or even thick clouds, and incidentally, they in fact didn't fly at night, presumably to avoid inadvertent flight into a cloud.

Answer to Question #5: The right choice is B. There must be enough material to absorb heat. A pointy object would be much more likely to get so hot it that would just burn off.

Answer to Question #6: It's A. Although the landing gear is armed at about 2000 feet (at which point the Orbiter is still following about a minus 17 to 19 degree glide slope, just before the pre-flare maneuver begins at about 1750 feet), the gear extension isn't actually initiated until it is within only a very few hundred feet of the runway (between 200 and 400 feet), and the orbiter is below about 310 knots. This can be as little as 15 seconds before touchdown. (Also, it takes about seven or eight seconds for the gear to fully extend!) Once the gear uplock hook disengage command is given, the main gear trucks are driven down by simple gravity-assisted free-fall, as well as springs, hydraulic actuators, and also aerodynamic forces. (The nose gear extension has an additional pyrotechnic assist.) The tires are rated to landing speeds up to about 225 knots, are inflated with nitrogen to about 300 psi, and the main tires can each support about 123,000 pounds (but only about 45,000 pounds for each nosewheel tire). Incidentally, the nosewheel tires have a life of two landings, and the main wheels, just one.

Answer to Question #7: B. In a way, lift is fairly intuitive, and in a way, it's not. Lift is a function of the shape of whatever airfoil you use of course, its angle of attack, velocity, as well as the air density and viscosity, and some of these properties are just lumped together in something called the coefficient of lift. The "lift equation" says that lift is equal to this lift coefficient, multiplied by the density of the air times half of the square of the velocity, in turn multiplied by the wing area. (The density times one-half of the velocity squared is sometimes called the dynamic pressure.) But since indicated airspeed is proportional to the *square root* of air density, and the Martian air is about one percent as thick as it is here on earth's surface, the fact that the wings would need to "see" the same indicated airspeed would mean that the true airspeed would have to be the square root of one one-hundredth, or about *ten times greater*! Since the speed of sound in a gas is mostly a function of temperature--it also depends on the specific heat of the gas--and the average temperature on Mars is, very roughly, about minus 55 degrees C, sound would travel about as fast there as it does up in the stratosphere here on earth, which is actually only about 15% slower...somewhere around 575 knots! That means that you'd be just about supersonic, yes. But...there is one thing that would be in your favor: gravity. The gravity on Mars is about one-third that of earth, which would decrease takeoff speeds to a bit below Mach 1. Don't expect to do turns about a point though, at least not one that is nearby, because at those speeds and with the same mass, your momentum will take you many miles before you even started turning. And landings would be pretty hairy, because with one-third of the weight (and one third of the traction) you would need a l-o-n-g runway to slow down.

Answer to Question #8: This is quite a striking photograph. In it, an F/A-18 Hornet assigned to Strike Fighter Squadron One Five One, USS Constellation battle group, is shown "breaking the sound barrier" in the skies over the Pacific Ocean on July 7, 1999. It's choice (C), all right. Under the right conditions of temperature and humidity (as well as altitude, wind speed, and the shape and trajectory of the plane) a "vapor cone" effect results. The shock wave is visible as a large cloud of condensation formed by the cooling of the air. A smaller shock wave can be seen forming on top of the canopy.

Chapter 7: Avionics and Navigation

There are an astonishing number of amazing things to learn about the origin, development, and implementation of the navigational equipment and procedures that pilots use every day to get from one place to another. While some of these may admittedly be historical curiosities, there are some lesser-known aspects of operating navigational equipment which are interesting, and which could prove useful.

Subject: Fly Runway WHAT?

Question 1: When a controller asks a pilot to "maintain runway heading", what does he REALLY mean?

A) It's an inside joke with controllers. It's like "We'll do lunch" in LA. It means "drop dead".
B) Very funny. No, he really wants you just to track the extended runway centerline, compensating for any wind drift.
C) Close, but no cigar. It means that you should maintain the compass alignment of the departure runway, regardless of where you wind up.
D) It is a little known fact, but you should actually follow the precise magnetic heading to which the runway number is an approximation. (For example, taking off from runway 14, the actual heading might be 137.)

Subject: DMEs

Question 2: Is there a reason why you should <u>ever</u> shut off your aircraft's DME (distance measuring equipment), if you're not using it?

A) You should never disable any navigational equipment, on principle (the principle being "if you've got it, use it").
B) Yes, though it's considered an unlikely scenario. Unlike the passive VORs, which operate in the VHF frequencies, DMEs are UHF, and the signal can travel up to 200 miles. More importantly, it's an active system, with the the initial signal being sent by the airborne unit, and even more critical is that each DME installation has its own "signature" for the signal and return pulses (which are actually each TWO pulses separated by an amount particular to THAT installation). When the number of airborne originations reaches a certain value (coincidentally, about 200), the ground DME component gets "saturated" and can go off-line. This is why you should turn it off when on the ground, or when you don't need it, particularly when you are in a busy terminal environment.
C) You only need DME above FL 240.
D) If you need to conserve power, such as when your alternator goes South, the DME draws a large amount of current, and it's always transmitting!

Save your power to key that mic and to operate your most critical nav receiver.

Subject: From the "Official Jocularity" Department

Question 3: Where will you hear the call signs "Cactus", "Tsunami", or "Cowboy" on ATC frequencies?

A) These are call signs used by the military for various training missions, so you probably won't hear them.
B) Nowhere. This is CB lingo, and its use is strongly discouraged by the FAA.
C) These are the assigned call signs (plus a following tail number) for some of the "fighter pilot for a day" flight schools.
D) You might if you were in Las Vegas, Phoenix, San Diego, or Austin,Texas. These are the official call signs of three different regional airlines and charters.

Subject: Uh Oh! What's THAT Mean?

Question 4: Nearing the end of your return flight home, as you're descending to enter the traffic pattern at your home airport, in addition to some nearby red and green lights, you happen to notice a blue flashing light above you. What does this mean?

A) You are in serious trouble. In addition to a flashing or occulting red beacon at the top of tall towers, a flashing blue light is used to indicate the point where guy wires are attached.
B) It's a blimp. Just don't climb.
C) Although it is done fairly rarely, whenever parachute jumps are made at night, the jump airplane must display this, in addition to other position and navigation lights.
D) Air-to-air refueling is being performed.

Subject: A Different Slant On Hand-held Nav/Coms

Question 5: Why might you want to pay particular attention to the orientation of the antenna of your hand-held nav/com?

A) Holding it at a 45 degree angle from the vertical when inside an airplane's cockpit minimizes signal attenuation, due to interference from the fuselage.
B) Always point the antenna towards the windshield to maximize exposure to the more radio-friendly (and less radiation-blocking) plexiglass. This maximizes transmission power and effective range.

C) Simple: We need our eyes in the cockpit, and waving pointy sticks around in a confined space is just a safety hazard! Remember what your mother said!
D) Communication antennas are vertically polarized (or more strictly speaking, oriented), whereas navigation antennas are horizontally polarized (oriented). What this means in plain English is: hold it up-and-down when you're talking, but sideways when you want to navigate using a VOR.

Subject: NOT A Good Sign!

Question 6: Near the end of a long flight on autopilot, above an overcast at 9,000 feet, at night, a C-182 pilot suddenly realized that he had fallen asleep, and worse, that he had mis-set his OBS with the wrong course. Not being able to contact Center on his current or previous frequency, he frantically cycled through all frequencies in the hope of reaching somebody--anybody. Just before resigning himself to making a mayday call on 121.5, he was elated to hear that 131.8 is quite busy. Actually, he was worse off than he could imagine. Why?

A) This is the frequency used by range officers in a Military Operations Area.
B) This is the air-to-air frequency often used by airliners over the North Atlantic.
C) He's flown South outside of Miami Center and into Cuban airspace.
D) This was the frequency used for chase planes during a Space Shuttle launch.

Subject: Hold On There!

Question 7: A pilot is about to shoot an instrument approach. He's flying inbound from a nearby VOR toward the Initial Approach Fix (IAF) on a radial which helps define the IAF, and which is located on the final approach course. There is no "NoPT" printed on this radial, but there is no procedure turn on the approach plate, only a holding pattern. Furthermore, the pilot is at the right altitude for the approach, has also been cleared for the approach, and the amount of turn needed for entry onto the final approach course is minimal; not only is it well under 90 degrees, but in fact it is under 20 degrees. The pilot is not under radar control.

Even though the (direct) entry into the hold would be dirt simple, is the pilot required to fly the hold before commencing the approach?

A) No. The Aeronautical Information Manual states that a procedure turn is required only when it is necessary to perform a course reversal in order to establish the aircraft inbound on an intermediate or final approach course. This is also understood to apply in the case of a holding pattern.
B) Yes, because the AIM also says (in paragraph 5-4-8A) that a procedure turn, or a hold *in lieu of* a procedure turn is a required maneuver.

C) There is no correct answer to this question, because what constitutes a course reversal is itself not defined in either CFR Title 14 *or* the AIM (although one would guess that semantically, anything greater than 90 degrees would automatically imply a course reversal by definition, anything less than 90 would not (and perhaps exactly 90 might be "undefined").
D) If Visual Meteorological Conditions prevail, or the heading change is under 45 degrees, no course reversal is required.

Subject: Now You Tell Me!

Question 8: Real-life scenario: At 6000 feet, while enroute on an IFR flight plan, in visual conditions, you're flying with passengers on what for them is their first flight in a general aviation aircraft. About three miles ahead you see a cloud deck, and as you approach, you become certain that the cloud tops are about 500 feet above you. Not wanting to make your passenger nervous, you request "higher" but the controller says he can't oblige. What is your most expedient option?

A) Request a heading that will avoid the clouds, followed by another to return on course.
B) Cancel IFR and either divert around the clouds or climb to 8500 feet.
C) Accept your fate and reassure your passengers that you have an instrument rating.
D) Request an "on top" climb.

Subject: Black Arts

Question 9: When was the first practical "black box" required aboard airliners?

A) 1903
B) 1925
C) 1934
D) 1958

Subject: ..--..

Question 10: If you hear a repeating "dit-dah-dah-dit" (the letter "P" in Morse code) on 121.5 MHz, what kind of device is transmitting the signal?

A) a 406 MHz Personal Locator Beacon (PLB)
B) Actually that's *two* letters, being the Morse code for the letters "A" (dit-dah) and "N" (dah-dit) from an outdated (but still operational) radio range.
C) Over-flying an outer marker beacon (75 MHz) will cause any VHF radio to pick up a harmonic of its 400 Hz audio tone on 121.5 MHz, which

pulses at 300 milliseconds and sounds like Morse code for the letter "P" (even though it isn't).

D) The "P" code from your GPS can be heard on your comm radio on 121.5, if the squelch is turned up fully.

Subject: Wilbur, Come Here. I Want You. Over...

Question 11: When was the first radio used aboard an airplane?

A) 1920
B) 1912
C) 1910
D) 1901

Subject: Getting the Red Carpet Pulled Out From Under You

Question 12: What IFR aircraft under ATC guidance in IMC (instrument meteorological conditions) may suddenly find that their services have been terminated, and that they have to fly a full instrument approach, rather than simply being vectored to the final approach course?

A) any aircraft
B) any aircraft, when radar coverage is no longer available
C) any aircraft except military aircraft on Department of Defense missions
D) any aircraft, *unless* the President of the United States is aboard

Subject: Polly Want A Cracker?

Question 13: How did the use of the term "squawk" originate, in reference to transponder codes?

A) It was originally an acronym, which stood for: Sequenced Query, Unknown Address, Wide-Key.
B) The modulated carrier sounded fairly disagreeable--the prototype equipment had an audio monitor patched into the test circuit--not unlike the screech of a very vocal tropical bird.
C) Irrespective of such remarkable mimics with bright plumage, the term was used in reference to another nickname given by the British during the development of this device during World War II.
D) It came from the loud feedback that would often leak into the audio of communication radios during the Korean War.

Subject: North by Northwest

Question 14: How long have directions been expressed in degrees clockwise from North?

A) only since 1947
B) since World War I
C) since the mid-1700s
D) only since the Pelopponesian Wars!

Subject: All In Knots

Question 15: What was the origin of the now universally applied unit of measuring speed, the knot, in determinations both aeronautical as well as nautical?

A) The term comes from the word *naught* (also spelled as *nought*), a now anachronistic word which meant zero. From the nautical (ouch) terminology of the mid-eighteenth century, "naught" originally meant 1000 fathoms (which of course had three such digits) in an hour's time.
B) piddle! A knot was a somewhat derisive term used to refer to a "chip log" which was actually a carved piece of wood that was also weighted so it floated low in the water somewhat like a buoy. A light line was attached to it, the chip log was heaved overboard, and the line was then played out as a ship was under way. It was literally knotted at regular intervals (in the case of the US Navy, 47 feet, three inches), and the number of knots that had run over the side during the interval after which a precisely timed sandglass was turned over and then run out (which in the Navy's case, was exactly 28 seconds) thus indicated the ship's speed. ($47^1/_4$ feet in 28 seconds is darn close to 6080 feet in 3600 seconds.)
C) The term comes from the age of steam power. Back in the days of coal bins and sweaty men with shovels, when the coal ran out, the next option to keep running was to tear out the internal planking of the crew's quarters, which was often made from knotty pine (hence the term). Although it originally had no fixed quantity, a ship's speed came to be associated with the number of revolutions per minute of the engines (originally determined when a ship was new, had a clean hull, and was running through calm seas).
D) A knot was a term used to refer to a "pressure log", essentially a tube bent at a right angle so that its open end was submerged and faced forward into the water (in much the same way that a pitot tube is configured to register dynamic air pressure). The height of a column of water in the tube's vertical segment gave an indication of forward speed. The connection between this metal tube and a knot of wood (though somewhat tenuous) came from the fact that a cylinder of wood, usually made from a knot, was used as a "cork" (into which was affixed a calibrated wire) and indicated the top of the water level in the pipe.

Subject: The First Aeronautical Charts

Question 16: The first aeronautical charts didn't appear until

A) 1834

B) 1888
C) 1909
D) 1917

Subject: Reinventing the Wheel

Question 17: In what year did the first "whiz wheel" flight computer appear?

A) 1881
B) 1903
C) 1910
D) 1917

Subject: From Whence Comes the Wind?

Question 18: If you knew only your true airspeed and your angle of drift from the wind, is it possible to compute the direction and velocity of the wind?

A) Yes.
B) Almost, but not quite. You would also need your heading.
C) No. Not only do you need heading information, but also your original course.
D) No. In addition to the above, you must know your actual ground speed.

Subject: The First Aeronautical Charts, Part II (U.S. Sectionals)

Question 19: US aeronautical charts known as sectionals have been around for a long time. *How* long?

A) over 30 years
B) over a half century
C) over the last 80 years
D) a few years shy of a full century

Avionics and Navigation: Answers

Answer to Question #1: The answer is really choice C.

Answer to Question #2: The correct answer is B. (That figures--it's the most long-winded...)

Answer to Question #3: D is the right choice. They are America West, Pacific Wings, and Austin Express, respectively.

Answer to Question #4: It's D. A blue light means that air-to-air refueling is taking place. (And if your radio could transmit on 266.500 MHz, you might try asking them yourself.) Actually there are various lighting assemblies used on the aerial refueling pod (such as red, green, and amber), as well as on the tanker aircraft itself.

Answer to Question #5: D. This also means (aside from the fact that you'd have to tilt your head at an awkward angle to do this) don't expect much navigation performance if you hook up your hand-held to the aircraft's external com antenna!

Answer to Question #6: The right answer is choice B.

Answer to Question #7: B, Your Honor, with an explanation: The Instrument Flying Handbook (FAA-H-8083-15) offers some guidance here. On page 8-10 it notes that during such an approach, the holding pattern must be followed, except when radar vectoring to the final approach course is provided or when "NoPT" is shown on the approach course. (The Handbook doesn't define course reversal, either, by the way.) Of course, at least in many of the busier areas in the country such as the northeast, where I fly, this may be academic as one receives radar vectors almost all the time. But, whew! Those procedures! gotta love 'em!

Answer to Question #8: choice D: Like "contact approach", "on top" is a magic word for special privileges (and the pilot has to request it). If you inform ATC that you are in "on top" conditions and ask for an "on top" climb to 8,500 feet, unless there are traffic separation issues, the controller will probably approve the request with instructions to climb as requested, report reaching on top, and/or to maintain on-top conditions at or above (or below) some particular altitude. When requesting "on top" one must remain in VFR conditions and accept the responsibility for traffic avoidance, but ATC is no longer bound by the separation rules that govern IFR aircraft, so they are free to let you climb where you want. You *do* have to use an appropriate VFR cruising altitude, based on your magnetic course (and it must also be above any minimum IFR enroute altitude) *and* you must comply with VFR cloud clearance criteria, *and* of course you have to maintain VFR conditions, but your route isn't compromised. Controllers will also still provide traffic advisories and safety alerts. (However, if you will encounter IMC at any time, you must inform ATC that you want to resume your IFR flight at a "hard" IFR altitude, whereupon you will be given a new clearance.) Note that you don't actually have to be trying to overfly a cloud layer in order to request this. You can also be between layers, or even in the clear wild blue.

Answer to Question #9: It's D. There was a huge increase in commercial air travel after WWII, but in the early 1950s, many tragic airplane crashes occurred, and with no surviving crewmembers (or even passengers) to explain what went wrong, investigators could only speculate. The solution was the FDR, or the flight data recorder. In the 1930s, an Imperial Airways' Handley-Page HP-42 airliner had a crude form of FDR, recording engine rpm, as did the Airspeed Ambassador in the 1950's. (Actually, if you want to get academic about it, the Wright Brothers pioneered the use of a device to record propeller rotations, so if you want to look at it that way, then that would push it back a ways.) The first requirement to have recorders on board certain aircraft was published by the Civil Aeronautics Administration (forerunner of the FAA) on August 1st, 1958. A similar rule was issued in the United Kingdom in 1960 (although there were some uses of them several years earlier). It said that all civilian passenger carrying aircraft over 20,000 pounds should carry a crash protected flight recorder. Flight data recorders that were introduced in the 1940's weren't much more than etching pens on steel foil (basically a strip chart) and could not meet the requirements for survival of the forces involved in an aircraft crash (nor the possible fire exposure) until 1958.

While he did not actually originate the idea, the inventor of the first practical FDR was Dr. David Warren of the Aeronautical Research Laboratories in Melbourne, Australia. His crash and fireproof device was able to record the flight crew's conversation along with a few instrument readings. It became known as the "black box" due to its concealment, both in terms of location and functionality. Invented in 1953 and in production by 1957, the first ones were painted bright red or orange to make them easier to find after a crash. The first such recorder was called the "ARL Flight Memory Unit", but they were not made compulsory in Australian planes until years later (although Australia claims they were the first country to mandate their use). In 1958, world aviation authorities approved minimum operating requiremens for FDRs. Cockpit voice recorders were mandated for all commercial operators in 1965.

David Warren was born in 1925 in northern Australia. A very bright boy, he attended boarding schools in Sydney for 12 years, but in 1934 when he was nine, his father was killed in one of Australia's earliest airline crashes. Later, he began building radios as a hobby but when the wartime ban on amateur radio squelched that, he turned to chemistry as a hobby and ultimately, as a career. He taught science for two years and then lectured in chemistry at Sydney University. At the start of the space age, he joined the newly-formed Woomera Rocket Range and was sent to England to perform research on rocket fuels at the University of London, where he earned his doctorate. Somewhat ironically (or rightfully), he was involved in investigating the crash of the world's first jet-powered aircraft, the Comet, in 1953. He argued that a cockpit voice recorder would be a useful means of solving otherwise unexplainable aircraft accidents. The idea initially received little interest, so he designed and build a prototype, to demonstrate the concept. The modern-day equivalent of this device is installed in passenger airlines, worldwide. They are made by several companies.

Dr. Warren, with his prototype

modern units

photographs of Australian scientist Dr. David Warren and flight data recorders
photo at left provided by the Defence Science and Technology Organisation (DSTO) - http://www.dsto.defence.gov.au
permission granted from Defence Science and Technology Organisation, Australia

To say that flight data recorders must be sturdy is an understatement. They must withstand a 3400-g deceleration in 6.5 milliseconds, penetration by a one-quarter inch hardened steel spike dropped from 10 feet with 500 pounds of force, a 5000-pound pressure against all six axis points for five minutes, 1100 deg C for 60 minutes (such as in a fuel fire), or 260 deg C for 10 hours (a baggage fire), immersion in sea water for 30 days at an equivalent depth of 20,000 feet, and a 48-hour immersion in aviation fuel, lubricating oil, hydraulic fluid, toilet flushing fluid, or fire extinguishing agents. Their batteries have a shelf life of six years. Their underwater locator beacon is designed to continue "pinging" at 37.5 kHz for 30 days at depths of up to 14,000 feet. CVRs and FDRs cost roughly between 10 to 15 thousand dollars each.

In the United States, a minimum of between 11 and 29 parameters are required to be stored and in 1996 the proposal was made to increase this range to 17-88, depending on aircraft type. Many of the available recorders exceed these requirements, their solid-state components storing over 700 parameters, which are typicaly storable for 25 hours before any crash, and help investigators to pinpoint the development of faults, as well as to help picture the aircraft's final moments. Modern CVR's are required to store the last 30 minutes of speech immediately preceding a crash (two hours, for solid state units). Both types of recorders are kept in the tail of an aircraft, because following any impact (which usually involves significant forward motion), the rear of the aircraft is both cushioned somewhat by the fuselage preceding it, as well as being decelerated to a slower speed, and thus recorders have a much greater chance at survival. Examples of some collected FDR data are time, altitude, airspeed, vertical and longitudinal accelerations, heading, times of each radio transmission, pitch and roll attitude, engine thrust, control and control surface positions. CVRs record ambient cockpit sounds (including the sounds of engines, flaps or landing gear deployment, and conversations between pilots and ATC). Any of these might provide vital clues about what caused of a crash or incident.

Incidentally, CVRs have a "bulk erase" button that pilots press after a safe landing. This was added at the request of the pilots union to protect their privacy. Typically, it is interlocked with the "squat switch" that detects weight on the landing gear, preventing accidental (or intentional) erasures in flight. There is also a "soft-crash" feature, where if a very hard landing is detected, it stops recording after 10 minutes. This is to prevent meaningful data from being overwritten, since the electrical system is still functional after such an "arrival".

There is also an interconnect between the FDR and the CVR. The FDR produces a chirp once every four seconds with an encoded time stamp. The audio of that chirp is superimposed on one of the CVR channels. When that channel is played back in the lab, the FDR timestamp can be recovered from the chirp. Of course, the FDR also records its timestamp in its own data. In this way, investigators can synchronize FDR and CVR events. More recent models and dual FDR/CVR recorders make this link digitally (and military fighter aircraft or helicopters use combined "one box" CVFDRs).

Answer to Question #10: Answer: It's a low-power homing signal, transmitted by the PLB to help searchers with 121.5 MHz direction finding equipment in locating the PLB--and its user. (Why the letter P? I don't know, but it was suggested by the National Telecommunications and Information Administration. It might just be because that's the first letter in "PLB".) Anyway, PLBs were authorized for use throughout the US under Part 95 of the FCC Rules beginning on July 1, 2003. (Canadian-approved PLBs are *already* in use in Alaska, under an FCC developmental license granted to the state.) As far as choice B, with the old four course radio range, first in use over seventy years ago, the user would hear *either* an "A" or an "N" *or* if he was "on the beam", a steady hum. Also, they used much lower AM radio frequencies, so you couldn't possibly hear anything on 121.5.

Choice C is nonsense, as you wouldn't get radio frequency interference on a frequency that isn't an integer multiple of 75 (which 121.5 isn't) and even if you somehow had crossed wires with your marker beacon receiver, you would hear dashes (a Morse code "T"), not short and long pulses. (The inner marker though, modulated at 1300 Hz, *does* use alternating dots and dashes, which one could say is either an "A" or an "N".) Choice D is equally goofy, as the P-code is binary; it's not an analog signal. It's also over one gigahertz, and so is not modulated on a VHF carrier! It's A, folks.

Answer to Question #11: We don't normally give much thought to the connection between radio signals and our flying safety (until the radios stop working). The significance of 1901 is that on December 12 of that year, Guglielmo Marconi successfully transmitted a radio message (not much of a message though, being just the letter "S" in Morse code) across the Atlantic Ocean. Just two years later, another great accomplishment occurred, that of the Wright Brothers' flight. In the years shortly after these accomplishments, an innovator with the (now) amusing name of Elmo Pickerill imagined being able to combine the two discoveries. Pickerill was a wireless operator for the Associated Press, United Press, and the Hearst Newspapers. After the turn of the century, he worked with Dr. Lee de Forest and Guglielmo Marconi to establish wireless stations throughout the country. In 1909, he asked Orville Wright about renting an airplane and pilot to fly him and his equipment, but when Wright told him that there wasn't an airplane that could lift that much, Pickerill said that *he* would learn to fly, and then carry his device (rather than a pilot), performing his tests as both pilot and wireless operator. Two months after he began taking flying lessons, he was a qualified pilot. On August 4, 1910, he made an historic flight from Mineola to Brooklyn. Flying at a thousand feet and using his "push-button" telegraph key, he made contact with three wireless stations aboard ships, two coastal stations, a portable station at Manhattan Beach, New York, and with a station in New York City. On that day, air-to-ground wireless communication became a reality for airplanes. (Also given credit for an early air-to-ground communication on August 27 of that year were J.A.D. McCurdy and "Casey" Baldwin, from a Curtiss biplane over Sheepshead Bay, NY.) The term "wireless" was officially changed to "radio" by the U.S. Navy in 1912, but the term stuck until about 1920, when the Marconi Wireless Telegraph Company of America went out of business and a new organization, Radio Corporation of America (RCA) began operating. Radio direction finding as well as communication equipment were used by Zeppelins even before World War I, in fact, <u>as early as 1912</u>. The best answer is choice C. I would be remiss incidentally, if I didn't also mention that radio messages *had* been transmitted and received by balloons in France, Belgium, and Germany even earlier than this (by 1908), but the question did say "an airplane"...

Answer to Question #12: Answer: For those who aren't familiar with instrument approach procedures, the deal is that, generally speaking, a pilot in instrument conditions flies to this fix or point in space called the Initial Approach Fix, and then goes outbound away from the airport for a couple of minutes, possibly descending to some lower intermediate altitude. He or she then makes a controlled course reversal, and flies in toward the airport, intercepting whatever electronic guidance is available there, descending to a minimum altitude or decision height. In the real world, the vast majority of the time in controlled airspace at least, pilots get help from ATC in the form of "radar vectors" to intercept that final approach course at an angle which is small enough to preclude their having to do all that course reversal and turn-around jazz, and it saves time (and money).

As for the answer, nope, it's not D; even the President isn't exempt. This has actually happened, too (on October 21, 1988, in one particular case). Frederick, Maryland, home of AOPA, is also the designated reliever airport whenever the ceiling and visibility are *so* bad than even the Marine One helicopter can't get the President into Camp David. In such cases, then Air Force One for example will do the ILS 23 into Frederick, and the President would continue the last dozen or so miles by car. Even though nearby Dulles airport has radar that can see a bit lower than the other nearby major airport, BWI, before the consolidated Potomac TRACON it was Baltimore Approach which was the controlling facility. (It may be due to the fact that Baltimore was designated as the controlling facility for FDK even before Dulles existed. This was a local political battle that raged for years.) As the story goes, Dulles radar controllers could see down just about into the pattern at KFDK, while Baltimore's radar wasn't reliable below about 4000 feet! (This has improved somewhat since then, thanks to ASR-9 radar.) But when you drop off the scope, forget radar vectors to the final approach course, even if your plane is sporting the Great Seal (or another non-designated facility can see you). You're on your own. The answer (finally) is B.

Answer to Question #13: The origins stretch back into aviation's earlier history, and as with many, this one goes back to the Second World War. After radar was developed, a method was needed to distinguish between friendly and unfriendly airplanes. The answer was transceivers that could respond to radar interrogations with a specific code. Allied aircraft could be identified by such unique codes while the enemy (for the time being) had no such codes. The Brits called the system "Parrot." The more common term was IFF, which stands for Identification (or by some accounts, Interrogative) Friend or Foe. The radar station would instruct a pilot to "squawk his Parrot" (that is, to a specific code). While Polly has flown elsewhere, that term remains today, meaning to set a code in our transponders. It's choice C.

Answer to Question #14: First off, I offer my apologies to the late Cary Grant, who starred in the movie along with Eva Marie Saint and the late James Mason, and the late Alfred Hitchcock, who made the 1959 adventure thriller, and the late (or almost late) anybody *else* who had anything to do with it.

There have historically been only a few ways by which direction was expressed in modern times, although all of them involved the compass rose. In marine navigation since the 1300s in the western world, there were the traditional cardinal "points of the compass" (of which there are at least 32 incidentally), which was the oldest system. In between, say North and West was Northwest, thus bisecting the four main "winds" or points of the compass rose into eight 45-degree slices. Then, further dividing these eight points, in between Northwest and West for example was West Northwest (WNW), and along with NNW, NNE, ENE, ESE, SSE, SSW, and WSW, the compass was thus further separated into 16 $22^{1}/_{2}$ degree wedges. Then, in between *those* and their less complicated neighbors lay (taking WNW again as an example) names like Northwest by West and West by North, which gave the 32 points of the compass, each $11^{1}/_{4}$ degrees apart. Even *that* wasn't enough for the more precise steering that became possible with steam-powered vessels, and then half and quarter points were added, further dividing up the compass rose into 128 skinny wedges, each one only $2^{13}/_{16}$ degrees wide. (They had names like "Northwest by West, 3/4 West". Clearly, that wouldn't work in today's National Airspace System, now *would* it?) The second system had the compass rose graduated in 360 divisions, but numbered from both North and South, increasing to 90 degrees at *both* East and West. It was the third system, just coming into use by various nations only in the late 19th century, which made the most sense. It involved the same 360 divisions, but counted them clockwise from North. The United States and Great Britain were *not among* those nations, by the way. *Only in 1916* did it drop the compass points (as other than additional decorative information) from its charts. It's B. Aye, aye!

Oh, and one more thing: There actually isn't *any* such point of the compass as "North by Northwest". They should have spent a little time looking at a Boy Scout handbook, or perhaps hired a consultant. There's Northwest by West (which equals 303 degrees, 45 minutes), and there's Northwest by North (326 degrees, 15 minutes). There *is* a "NNW" between due North and northwest, but there is *no* intermediate "by" for any of the 16 "higher level" points of the compass (those appearing in bold type, below). Here are all 32 of the most well known compass points, converted to degrees clockwise from North:

NORTH	**0°**
N by E	11° 15'
NNE	**22° 30'**
NE by N	33° 45'
NE	**45°**
NE by E	56° 15'
ENE	**67° 30'**
E by N	78° 45'
EAST	**90°**
E by S	101° 15'
ESE	**112° 30'**
SE by E	123° 45'
SE	**135°**
SE by S	146° 15'
SSE	**157° 30'**
S by E	168° 45'
SOUTH	**180°**
S by W	191° 15'
SSW	**202° 30'**
SW. by S	213° 45'
SW	**225°**
SW by W	236° 15'
WSW	**247° 30'**
W by S	258° 45'
WEST	**270°**
W by N	281° 15'
WNW	**292° 30'**
NW by W	303° 45'
NW	**315°**
NW by N	326° 15'
NNW	**337° 30'**
N by W	348° 45'
NORTH	360°

table of Conversion of Compass Points to Degrees, by the author

Answer to Question #15: The "*naught*-ical" etymology given in choice A is entirely fictional. Choice B is the correct answer. Several different types of speed measuring devices were in use before the pitot tube became the world standard. There actually *is* some historical truth in the correlation between engine revolutions and forward speed made good mentioned in choice C, and the pressure log mentioned in choice D also really existed (although that stuff about floats fashioned from knots was just bilge water). Other early means of measuring speed included hinged pressure plates, and even airborne anemometers. The nautical precedent favoring knots as a unit of speed (metrification knotwithstanding) certainly *would* have been hard to ignore!

Answer to Question #16: There were indeed maps available for aeronauts of the 18th and 19th centuries, but not one of them had been prepared for aerial navigation. Towns were noted in terms of population, not by dimensions or outlines. Railroad maps proved useful for balloonists, although the paths and precise directions of railroad tracks, and at what angles they crossed roads as well as each other, were far from realistic. The first published call for aeronautical charts was made by a Prussian officer and balloonist named Hermann Moedebeck, in 1888. He pointed out that it was difficult to determine safe landing sites from aloft, and suggested that distinctive symbols be used to show such locations. Actual production of such charts however, was still over 20 years away. In 1900, a subcommission of the Second International Congress of Aeronautics did consider the problem of locating one's position from a balloon, but the problem wasn't really addressed. In 1906, Moedebeck once again advised that a better means of determining position by representing useful landmarks was needed, and even suggested governmental support. In 1907, the Third Congress of the International Aeronautical Federation (a rival organization) formed an international commission for aeronautical charts, and Moedebeck was appointed as its first chairman. As a prototype he chose a topographical chart being produced by the Royal Prussian Land Survey, which covered middle Europe on a 1:300,000 scale (the "Cologne sheet"), on which he over-printed various aeronautical symbols. It was shown at an exposition in Frankfurt in 1909. (Yes, the answer is choice C.) Also in 1909, a commercial firm sold an aeronautical chart of Paris; in 1910 the Aero Club of France organized a cartographic commission, and by 1911 it had received a subsidy from the Ministry of Public Works. In the fall of 1911, the first air chart ever made in the US appeared. It was a chart of western Long Island, made for the Aero Club of America.

Answer to Question #17: Especially because the first aircraft flew so slowly, interest in quantifying the effects of wind on the path of an airplane or airship was more than incidental. Very quickly, the more mathematically inclined realized the value in vector diagrams and trigonometry. By using vectors to represent the respective directions and velocities of both wind and aircraft, connected head-to-tail as arrows representing a convenient interval of time, the resultant side of such a completed wind triangle revealed the path it would most likely take over the earth's surface. This was discussed in popular aviation magazines as early as 1910, as was a description of a "compass for aerial navigation" (in that same year). Loxodromic curves may be mathematically elegant, and arriving at one's destination pointed right into the wind may seem convenient, but such a course is inefficient. Pilots knew that "overcorrecting" into the wind to "kill the drift", so that a given landmark always remained at the same relative bearing (until overflown, of course) was the best way to fly. But not all flights were over land, for short distances, or in fair weather. In such cases, one needed the right heading to which the "steering compass" or directional gyro would help guide an aircraft most directly to wherever it was going.

It wasn't until World War I that the dead reckoning computer came into widespread use. (Until that time, aviators had precious little advance information of the winds aloft, but during the war, weather stations were able to provide informed estimates, at least.) The first such devices took the form of adjustable rulers, which were not so easy to use while also manipulating a chart in a cockpit. The second form was a rotating gridded circle, called a Course and Distance Calculator by the RAF, and was derived from a naval instrument. Having a graph and two movable arms, positioned to represent the known values of the wind triangle and clamped in place with an adjustable screw, the user could then simply read off the unknown component against the graph. A circular slide rule on its periphery complimented its functionality. The correct answer is choice C.

Answer to Question #18: Solving a navigation problem like this usually involves (and depends upon) one's first knowing true airspeed, ground speed, course, and heading. That's right out of private pilot training. Aha, but there's a trick here. The correct answer isn't a simple "yes"--scratch choice A--but you actually don't need to know original course, *nor* your ground speed. (The answer is choice B.) How could that be? By using two wind triangles, making two successive runs in two different directions, and then pooling the data, as shown in the diagram below, the wind vector could be taken in common and solved graphically. Obviously such consecutive runs should be made at the same altitude, in the same area, and as close together in time as possible. (A device, later called a "wind-gauge bearing plate," was first fashioned to solve such a wind star by a lieutenant colonel in the Italian Air Force.) By using the two wind triangles together and having the wind vector in common, both triangles can be solved. Once the headings, airspeed (presumably held constant), and drift angles are known, the ground track can be determined (using heading, plus drift). The intersection of the two heading lines connected in turn to the intersection of the two drift lines would reveal the wind velocity (that is, both its direction as well as its speed).

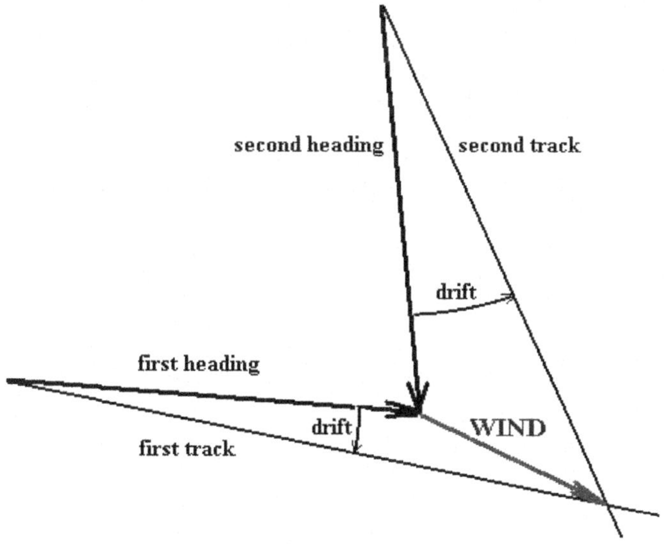

drawing of wind triangles, by author

Answer to Question #19: The need for aerial mapping of the U.S. was made quite obvious when airmail flights began during the last few months of World War I (first by the U.S. Army, then the Post Office). Pilots made do with whatever maps they could find (or in the case of enterprising pilots like Elrey Jeppesen, compile, from their own experience). The U.S. Army Air Service began photographing all routes to aid pilots in their orientation, and commercial firms such as Rand McNally produced "Air Trails" maps as early as 1928. When the US Government began printing aerial maps, at first they took the form of strip charts of principal air routes, similar to an AAA "Trip Tik" and were in use as early as 1923. In 1926, the Navy's Hydrographic Office printed coastal charts of North and Central America. Upon passage of the Air Commerce Act in 1926, placing civil aviation under the Commerce Department, the Coast and Geodetic Survey produced additional aerial maps. It reached the point where the country was covered somewhat haphazardly by different types of charts from various agencies; some areas of the country received no coverage whatsoever, while others were mapped multiple times in different ways. Finally, a committee representing these agencies met in 1929 to resolve the issue. They found that more flying was done off the airways, and not on them, and also that for military purposes, the country needed something more all-inclusive. The new chart, called a sectional chart, was a conic projection on a 1:500,000 scale, and most of them each covered six degrees of longitude (about that of today's sectionals) and two degrees of latitude, which is half that of current sectionals. (That' s not to be confused with the term used in surveying, where a "section" is the unit or subdivision of a township with one mile square rectangular boundaries, containing 640 acres.) The first sectional charts appeared in 1931 (so it's choice C). They were initially issued every two months, and there were roughly 87 of them (more than twice the present number). A more recent historical note is that on October 1 2000, the National Aeronautical Charting Office (NACO) and its 300-plus employees became a part of the Federal Aviation Administration, Aviation System Standards, as a result of an agency transfer of the Office of Aeronautical Charting and Cartography from the National Oceanic and Atmospheric Administration (NOAA).

Chapter 8: Helicopters

I'll be the first to admit that I've always been smitten by these things. Even less intuitively capable of thwarting the ties of gravity than those aircraft whose wings *don't* swing around, helicopters can't help but garner glances from even the most disinterested onlooker. For many of us, especially the hopeless cases such as myself, it's blind, unrequited love. Some may find such obsessions tiresome however, and to keep things on an even keel, I've restrained myself somewhat.

Subject: On Dynamic Rollover Angle In R-22's:

Question 1: How precipitous a hill can a Robinson R-22 helicopter land on, sideways along the slope, without rolling over and balling up? What are those angles; first, with full opposite cyclic, AND second with no compensation at all?

A) 35 degrees, 25 degrees
B) 42 degrees, 29 degrees
C) 22 degrees, 14 degrees
D) 15 degrees, 6 degrees

Subject: On Autorotation In A Robinson R-22:

Question 2: How long do you have to enter autorotation after an engine failure in an R-22?

A) a little over one second, or you're history
B) just under four seconds at minimum power speed of 60 knots
C) six seconds at 60 knots, eight at 85 knots
D) Hey, no sweat! There's a HUGE moment of inertia swinging around up there at a few hundred miles an hour, and you have about 10 seconds while you slowly and gently sink back to earth.

Subject: Dubious Parallels

Question 3: What do a toilet and the gearbox of an R-22 have in common?

A) The shape of the gearbox, viewed from the side, does tend to have a familiar silhouette...
B) The gear ratio is 47:11, which coincidentally is the number of the famous perfumed water that Napoleon was said to have used: Koinisch Wasser 4711.
C) They both use a hydraulic jump to aid fluid flow within their respective internal constrictions.

D) You can see the contents of both. In the case of the gearbox, there's actually a window to view the fluid level. Of course, in the other case, you might not want to discuss it.

Subject: Everybody, Get Outta the Way!

Question 4: What is the average area commonly recognized to be necessary when new students are learning to hover a helicopter?

A) Generally, with the instructor poised to take the controls if (when) necessary, a square open space of between a half-acre to one acre is usually enough.
B) It depends on the type of helicopter, and there is no fixed size. Smaller piston helicopters such as the Robinson R-22 are much more dynamically unstable than some larger turbine helicopters, such as a Bell 407. As a result, the area given in choice (A) usually becomes somewhat smaller, using larger machines. Also, there are some people who, uh...well, see the next choice.
C) New Mexico
D) at least four acres, to be absolutely safe, given varying instructor response times

Subject: No Strings Attached

Question 5: The connection between the helicopter and concert piano is that:

A) the auditory alerts for the Army's Comanche helicopter were played on a grand piano.
B) someone cut a grand piano loose at 3000 feet from a cargo helicopter near Seattle, prior to a concert by Country Joe and the Fish in 1968 (and again for an MTV stunt out in California City in the fall of 2000, with pretty much the same results).
C) if it weren't for Sergei Rachmaninoff, Sikorsky Helicopters probably wouldn't exist.
D) A and B

Subject: The First *Controllable* Rotorcraft

Question 6: In 1907, in France, the European cradle of aviation, the world's first rotorcraft lifted itself straight up. But the first vertical flights of Breguet and Richet's gyroplane (steadied by four men) and then Paul Cornu's machine a few months later (which was more stable, and which rose only a very short distance), were not truly controllable. Aside from the fact that the first truly flyable rotorcraft was not a true helicopter but an autogyro, who designed and flew the first rotorcraft with the ability to fly forwards, backwards, and sideways by means of a "cyclic" control?

A) Louis (and Jacques) Breguet
B) Igor Ivanovich Sikorsky
C) Raul de Pescara
D) Heinrich Focke
E) Thomas Edison

Subject: World's Fastest Helicopter

Question 7: The world's fastest helicopter could travel at

A) 150 knots
B) 216 knots
C) 263 knots
D) 630 knots

Subject: Starring Roles

Question 8: In July 2001, 13-year old Cody Clawson was hiking with his Boy Scout troop in Yellowstone National Park, but somehow, he got separated from the group. Even though authorities enlisted help from rescue teams in Idaho and Wyoming, Cody wound up spending a cold, wet, and lonely night in a rocky canyon, wondering if he would survive. The next morning, he heard a helicopter flying over a nearby ridge, and the Bell 407 pilot spotted Cody, landed nearby, and flew him back to safety. Who was that pilot?

A) Clint Eastwood
B) Harrison Ford
C) Cliff Robertson
D) John Travolta

Helicopters:
Answers

Answer to Question #1: It's D: 15 degrees, with opposite cyclic; a measly 6 without! But if it's shut down and plunked on a hillside with a crane somehow, it is a whopping 42 degrees! (That's independent of cyclic, of course. Being shut down, there's no collective raising the effective center of gravity. It's called the "static rollover" angle. Who would do such a thing is a question in itself.)

Answer to Question #2: It's A: Jack be nimble, Jack be quick, or Jack be day-ed. Robinsons are great, the engines are de-rated and almost never fail, but you can get carb ice...And unfortunately, they've got very LOW rotor ineria. Once you get below 70 to 80% rpm (altitude-dependent), it's over. And the rate of decay below that is much more rapid.

Answer to Question #3: It's B: Or, maybe B and D. (There really is a window.)

Answer to Question #4: It's B. Choice (D) isn't at all unwise, either. (This isn't really a "fact", I know. It was more an attempt at humor.)

Answer to Question #5: It's C. Igor Sikorsky was born in Russia in 1889, built many airplanes for the Czar, became fairly wealthy, and built the world's first four-engined airplane. But when he fled the Russian Revolution, he arrived flat broke in America. Six years later in 1923, with $800 in savings, he started the Sikorsky Aero Engineering Corporation. It was a rather humble emigre operation, strategically located on a friend's Long Island chicken farm. There were some twenty mechanics, several of whom were quite illustrious refugees, calling each other Baron, Count, or General, because they *had* been (although today that might conjure visions of asylum inmates). They made tools, as well as airplane parts, out of anything handy; their first project was a twin-engine passenger plane, the 29A. (It was Sikorsky's 29th design, and the A simply stood for America.)

But you might say that the famed composer and pianist Sergei Rachmaninoff was instrumental (pardon my highbrow guffaw) in getting helicopters off the ground. Besides being a very tall man, and having phenomenally large hands--he could play a 13th on a piano keyboard—Rachmaninoff had another kind of reach: he kept a watchful eye out for his fellow Russians. He had heard about Sikorsky, and on the very day that he first came to see the "factory" he gave Sikorsky a check for $5000 (about $100,000 today) with (dare I say it) no strings attached. This enabled Sikorsky to relocate to a hangar at Roosevelt Field, and rescued his teetering company from the brink of oblivion. For that, Sikorsky made Rachmaninoff the Vice President of his company, and the two went on to become close personal friends. Fittingly, Sikorsky's 29A made its first profit ferrying two pianos from New York to Washington. Not many years after Rachmaninoff's generous assistance, Sikorsky sent Rachmaninoff a check for that $5000, plus interest.

Answer to Question #6: A regular United Nations to choose from here, ain't it? The answer is C. In case you are unaware, a cyclic control varies the pitch of individual rotor blades over specific sectors along their path of rotation, as opposed to a simple collective pitch control of all blades at once (like, um, in an airplane). Juan de la Cierva is often credited with creating the first successful helicopter, which was actually an autogyro, where forward thrust was provided by a conventional airplane propeller and lift was provided through rotor blades which would have otherwise simply been in a constant state of autorotation. (The Russian inventor Boris Yuriev (also written as Yur'ev) proposed the idea of using an anti-torque rotor, as early as 1912.) Cierva's C.8 Mk IV made the first crossing of the English Channel by a rotary-wing aircraft in September, 1928. Heinrich Focke's Fa-61, flown by the well-known Hanna Reitsch in June, 1936 was an early rotorcraft, which looked like an autogyro, with twin rotors on outriggers on either side of what looked like an airplane fuselage (however the engine actually drove the rotors as well as a fron-mounted tractor propeller). Igor Sikorsky's VS-300 was the first successful US-built helicopter, in September 1939 (and it really looked like a helicopter). But it was an Argentinian engineer, the Marquis Raul Pateras de Pescara (who at the time was living in Barcelona) who first figured how to build a vertical flight machine that allowed the pilot to vary the pitch of the main rotor blades. Why is that so important? Due to the dissymmetry in aerodynamic loads between the sides where the rotors are advancing into the direction of motion, and retreating away from it on the other side, a means of changing the pitch on the blades during their rotation is needed. Also, by selectively varying their pitch even in the absence of forward motion, any desired lateral motion can be induced. To be fair, there *were* others who realized the necessity of having a means of effecting a cyclic variance in blade pitch, such as Yur'ev, as well as an Italian named Gaetano Crocco, in 1906, and also a Danish aviation pioneer named Jen Ellehammer. After moving to France in 1922, de Pescara's third design the following year incorporated the first credible use of cyclic and collective pitch controls (although it was based on wing warping). Incidentally, he also proved that a helicopter could still autorotate and land safely in the event of an engine failure. The world's first "true" helicopter is credited to Louis Breguet. His coaxial rotor *Laboratoire* was the first to use articulated blades to effect inputs from a cyclic control. It first flew in June 1935. Edison really did experiment with helicopter models, too (in the 1880s).

the de Pescara helicopter, ca. 1923

Answer to Question #7: Airplanes stall when they get too slow. With helicopters, it's just the opposite; they can actually stall because they are going too fast! Why? The answer can be inferred from the name of the phenomenon: retreating blade stall. All helicopter manufacturers must establish maximum airspeed limits that are below this threshold. For example, the rotor tip speed in an Robinson R22 at 104% rpm, or the "top of the green" on the tachometer, is somewhere around 370 knots. The helicopter's "never exceed" speed is just a bit over 100 knots (102 knots, at sea level). Even with the greater angle of attack that is allowed by the flapping of its under-slung and teetering pair of blades (or about the flapping hinge, on articulated rotor systems), there still comes a point where the relative motion of the "retreating" blade is below that needed to remain level (even with the blade on the retreating side having a greater angle of attack), and this dissymmetry of lift would cause an unfortunate helicopter to both pitch *and* roll over. This is *not* a good thing. (The "roll" part comes from that dissymmetry; the "pitch" is due to gyroscopic precession.) Helicopter manufacturers publish charts or graphs showing Vne decrease with altitude. Retreating blade stall is aggravated by high weight, low rotor rpm, turbulence, or abrupt maneuvering, as well as (of course) excessively high forward airspeed. Unlike airplane pilots, helicopter pilots often think in terms of their "percent rpm". At "full" rpm—with the governor turned on in an R22 for example, that's approximately 104%—as well as in forward flight right at Vne, the percent drop in tip speed on the retreating blade's side would be about 70%. (Note that this is at the faster-moving tip; it's much worse towards the middle.) This just happens to correspond to the percent rpm below which a helicopter experiences what is known as catastrophic rotor stall. Beyond that point, there is no recovery, even by fully lowering the collective. The rotor blades rapidly slow to a near stop, and the helicopter falls straight down, right out of the sky.

Although the Westland Lynx (choice B) can hit over 200 knots, the Lockheed XH-51A could top 250 (263, or just over 300 mph). In addition to the engine powering the rotors, it was also powered by a single J60-P-2 jet engine mounted on a stubby left wing. A number of helicopters can easily top 150 knots (choice A), but nothing with rotating wings can get anywhere *near* 630 knots. (Even the V-22 Osprey tiltrotor has a top speed that is, at most, about half that.) The right one is choice C.

photograph of the Lockheed XH-51A helicopter
from S. Harding "U.S.Army Aircraft Since 1947"

Answer to Question #8: It was B. It was Harrison Ford, who has a home in Jackson, Wyoming. All are pilots. Eastwood also flies helicopters (but not airplanes).

Chapter 9: Accomplishments of Questionable Merit

Aviation has traditionally been the province of those who don't exactly take a lackadaisical approach to living. What more perfect arena, other than show business, for someone who wants to step right up and remind the rest of us that we only come this way but once. Some of these folks don't just seize the day; they sneak up from behind, grab any unsuspecting opportunities by the horns, and then hang on for dear life. The rest of us can at least watch.

Subject: That First Step Is A Doozy!

Question 1: True or false: It is possible for a parachutist in free-fall to achieve a great enough velocity to break the sound barrier.

Subject: Getting Jumpy

Question 2: The world's record for the most parachute jumps (by a single person, now) in a 24-hour period is:

A) 65
B) 179
C) 350
D) 500

Subject: He Didn't Really Do That, Did He?

Question 3: America's highest scoring ace during World War II was Richard Bong. Flying the P-38 Lightning in the Pacific, Major Richard Bong was credited with 40 kills. A skilled flyer (to say the least), he was noted for his silent approaches with both engines feathered. As he swooped over the field, he would loop his P-38 and land. What else did he do?

A) In 1942, he looped the Golden Gate Bridge (in a P-38, of course).
B) On August 6, 1945 (the day the atomic bomb was dropped on Hiroshima) he was testing the new P-80 jet fighter, which stalled on takeoff. He bailed out, but too low.
C) just A
D) A and B (unfortunately)

Subject: Grand Entrances

Question 4: One of the first passenger jets to enter commercial service was the venerable Boeing 707. What did test pilot Tex Johnston do with a Boeing 707 prototype in front of about 200,000 people, at Seattle's annual Seafair festival in 1955?

A) He broke the sound barrier.
B) He flew in so low over Puget Sound that he generated a rooster tail.
C) He rolled it.
D) knife-edge flight

Subject: Air Bowling In the Faulklands

Question 5: True or False: The Audubon Society reported that Royal Air Force pilots stationed on the Falkland Islands search out a beach where the penguins are gathered and fly slowly along it at the water's edge. Perhaps ten thousand penguins turn their heads in unison watching the planes go by, and when the pilots turn around and fly back, the birds turn their heads in the opposite direction, like spectators at a slow-motion tennis match. Then, the pilots fly out to sea and directly to the penguin colony and overfly it. Heads go up, up, up, and ten thousand penguins fall over gently onto their backs.

Subject: Chasing the Blues Away

Question 6: Which extreme measures have actually been taken, with regularity, to dispel fog at airports?

A) wind tunnel fans (to blow it away)
B) sawdust AND fans (to dry it up and provide condensation nuclei)
C) heat (to lower the relative humidity)
D) high explosives

Subject: Turn That Crank; Steer That Rudder!

Question 7: The first time a propeller was used on a full-sized aircraft was in the year:

A) 1784
B) 1852
C) 1872
D) 1884

Subject: Pointy Pontoons

Question 8: True or False: A seaplane has been built which exceeded Mach 1

Subject: One Heck Of A Piñata

Question 9: True or False: Someone has built a passenger-carrying heavier-than-air aircraft that was made entirely from paper.

Subject: Talk About A Multi-engine Rating!

Question 10: The greatest number of engines used on any production airplane was:

A) Howard Hughes' HK-1 Hercules, N37602, a.k.a. the "Spruce Goose" (even though it was 95% birch) had **eight** Pratt & Whitney 3000-hp engines. (When the industrialist Henry Kaiser backed out, Hughes removed the "K" and changed the designator to H-4.)
B) The ca. 1950 Consolidated B-36D bomber had **ten** engines (six 3500-hp Pratt & Whitney R-4360 radial engines, and four General Electric J-47 jet engines).
C) Dr. Claudius Dornier's 1929 Do.X was powered by **twelve** 525-hp Bristol-Jupiter engines mounted in six tandem sets across the upper surface of the wings. (However, even when the engines were replaced with 610-hp Curtiss Conqueror engines, it could not fly any higher than 1650 feet, but that was then considered a minimal and acceptable safe altitude for trans-oceanic flight.)
D) Impressed by the Dornier Do.X, Norman Bel Geddes, a well-known American industrial designer and a German aeronautical engineer named Otto Koller, convinced that size was the key to safety and stately comfort, designed Airliner #4, having a 528 ft wingspan, nine decks, two elevators, ample accomodations for over 450 passengers, and a crew of 155. The hulls contained kitchens, crew quarters, lifeboats, and two aircraft which could be launched in flight. There were also well-appointed staterooms, a gymnasium, dining room with dance floor, promenade decks, verandas, and a bar. To get all this in the air, there were **twenty** 1900-hp internal combustion engines, although only 12 were needed to fly at cruising speed. There were also **six additional** 1900-hp engines in reserve, for a total of **twenty-six**!

Subject: Wings: the More, the Merrier

Question 11: Aside from the extreme example of the steam-powered "multiplane" designs of Horatio Phillips, which resembled a set of venetian blinds attempting flight, the greatest number of wings sported by any airplane of comparatively more realistic design was in the era shortly before, during, and after World War I. What was the second most, shall we say, ambitious?

A) The Fokker and Sopwith triplanes flew fairly well. Actually, there were other less well-known Sopwith triplanes such as the Snark and Rhino, as well as famous failures like the monstrous Tarrant Tabor, and quite a few

French designs like the Labourdette-Halbron, the Morane-Saulnier TRK, and the Astoux-Vedrines. The Germans had the Albatros Dr1 and Dr2, the Siemens-Schuckert DDr1, the Sablatnig SF 4, and the the Pfalz Dr 1 and Dr2 (to name just a few). And let's not forget the Italians. They had the Breda-Pensuti, the Caproni CA 40, 41, 42, 43... The Russians were perhaps more reserved. They had only the Bezobrazov.

B) Just three? Nope. Under the impression that "more is better" there were many quadriplanes. The Wright Type 4, the Armstrong-Whitworth FK 10, the Supermarine Nighthawk, and the Pemberton-Billing P.B.29 were real. So were the the Friedrichafen Quadruplane, and the the Naglo Quadruplane.
C) Hah! We're just getting warmed up, here! How about a V-8? No, not tomato juice, the The Fokker V.8 quintuplane. Yes, we're going for five.
D) About the only example having six wings that I could find was the Avro Bullseye...but...how about seven? The Johns Multiplane had seven wings. And it never flew.
E) It gets worse.

Subject: Icarus II: Falling With Style

Question 12: True Or False: Using wings no wider than a person is tall, it is possible for a human being of normal proportions to descend in free fall with a glide ratio that approaches that of some aircraft.

Subject: 5280

Question 13: The first candidate for membership in the mile-high club was:

A) Howard Hughes, in 1937
B) Nelson Rockefeller, in 1979
C) Lawrence Sperry, in 1916
D) Rudolph Valentino, in 1925

Subject: Flying Blind (and Flying Safe?)

Question 14: True or false: a blind person has successfully piloted an aircraft (an *airplane*) through takeoff, cruise flight, and landing.

Subject: They Shoot Horses, Don't They?

Question 15: The longest continuous flight ever made was

A) two hours and 13 minutes shy of 20 straight days
B) nine days, four minutes in the Voyager aircraft's world-circling non-stop (and non-refueled) flight between December 14 and December 23, 1987

C) an unbelievable 64 days, 22 hours, 19 minutes, and five seconds, in a Cessna 172, entirely in the vicinity of McCarran Airfield in Las Vegas, Nevada, between December 4, 1958 and February 7, 1959

D) You're not going to believe this one. In 1976, John Ewing and Blair Alberson made a non-stop flight lasting eleven months, 29 days, 23 hours, and (the records are hazy on this part) about 55 minutes, also in a Cessna 172 (highly modified to allow in-flight refueling by another C172). They didn't quite make a full year because of an engine failure. (They also had an STC to allow in-flight oil changes, as well as meal delivery and, um, comfort facilities.)

Accomplishments of Questionable Merit: Answers

Answer to Question #1: TRUE. However, you would need a helium balloon, and you would have to get a mile or two above 100,000 feet. The air is thin enough up there so that terminal velocity would actually be greater than Mach 1. (Also helpful is that the speed of sound becomes about 100 mph lower in the thinner air.) By the time such a human projectile reached around 85,000 feet, he (or in the case of someone like Cheryl Stearns, she) would for a few seconds more, be traveling in excess of Mach 1.

Answer to Question #2: The answer is D. The jumps were made from 2100 feet, on average, once every two minutes and 40 seconds.

Answer to Question #3: The answer is again, D ...unfortunately.

Answer to Question #4: The answer is C. He actually did it twice; the first time was on the way there, "to make sure it was okay".

Answer to Question #5: It's false. The attribution is urban legend. This phenomenon was first reported by Royal Air Force pilots who flew over the Falklands during the 1982 war with Argentina. Although it's easy to anthropomorphize penguins, they actually hate the sound made by airplanes and are known to scatter whenever one approaches. Late in 2000, British Antarctic Survey researcher Dr. Richard Stone spent one month aboard HMS Endurance studying the "falling penguin" phenomenon. The Associated Press reported Dr. Stone's findings: When an aircraft flies overhead, they do not topple over like dominoes, as some Royal Air Force pilots have reported. "Not one king penguin fell over when the helicopters came over Antarctic Bay," said Dr. Stone. "As the aircraft approached, the birds went quiet and stopped calling to each other, and adolescent birds that were not associated with nests began walking away from the noise."

Answer to Question #6: The correct choice is C. During WWII, in an effort to find ways of ensuring that bomber crews returning to fog-plagued England would make their home fields, research scientists at another location having equally persistent fog, Arcata, California, explored a very unusual technique to dissipate fog. Burning 60-octane gas under pressure, as well as from trenches along the sides of the runway, they created walls of flame 15 feet high. When it was "W0X0F" (the old pre-METAR surface aviation observation reporting sequence denoting "indefinite ceiling, sky obscured, visibility zero, fog") they turned up the heat, which brought ceilings up to the needed 200 feet.

Answer to Question #7: The world's first dirigible is credited to a Frenchman, Henri Giffard, who used a steam-driven (power output: 3 HP) propeller beneath his 143-foot long cigar-shaped balloon filled with coal gas to arise from the Paris Hippodrome on September 24, 1852. Although he did travel about 17 miles to Trappes at an average speed of about five or six miles an hour, his 11-foot three-bladed propeller (which turned at about 110 rpm) wasn't the first. (There was also a triangular sail fastened to the end of a pole, which he used as a rudder.) So it's not choice B. Nor was the dirigible that was first flown in 1872 (choice C) in Vienna Austria, by Paul Haenlein. His was the first dirigible powered by an internal combustion engine (four cylinders, 2.8 HP), which consumed gas from its own envelope and drove a 15-foot propeller at about 40 rpm. And it definitely wasn't the La France, flown on August 9, 1884 by Captain Charles Renard and Lieutenant Arthur Krebs from the French Corps of Engineers (choice D), which was however the world's first *fully-controllable* dirigible. At 166 feet long, powered by a nine horsepower electric motor which turned a 23-foot four-bladed wooden propeller, it flew at nearly 15 mph. (There was also an effort in 1882 by Gaston and Tissandier using a 92-foot long airship powered by a 1.3 HP electric motor, but although it flew a large semicircle at 7 mph, it wasn't very successful and barely made headway against the wind. No, the first time a propeller was used in any aircraft was way back in 1784 (choice A). The arrival of the free balloon the year before was a great leap in terms of man's attempts to escape the ties of gravity, but as a means of practical travel, it could scarcely be called predictable. Although many years would pass before realistic progress was made to steer and propel lighter-than-air machines at speeds greater than, and in directions other than, that of the wind, on October 16 of that year, Jean-Pierre Blanchard used a small hand-cranked six-blade propeller attached to the basket of his balloon. Note that I didn't say that the propeller had to be effectual. It wasn't. (The first design of an elongated and theoretically steerable balloon was published by the Frenchman Jean-Baptiste Marie Meusnier in that same year, as well.) Blanchard is also credited with having made the first manned free balloon flight in America, in Philadelphia, on January 9, 1793 (although a tethered ascent had been made in Baltimore, nine years earlier). In January of the following year, 1785, Blanchard and Dr. John Jeffries crossed the English Channel in a hydrogen balloon equipped with aerial oars and a hand-cranked propeller (neither of which had any effect on the flight whatsoever). After rising to about 2000 feet, they released some of the gas to prevent their envelope from splitting. They belatedly realized however that they had released a bit too much, and they spent the remainder of their voyage from Dover to Calais wildly tossing all excess ballast overboard, followed by their oars and propeller, then anything that wasn't fastened down, and finally, most of their clothing!

Answer to Question #8: It's true! (And no, it didn't have pontoons.) It was called the Sea Dart, it was built by Convair, and was also known as the XF2Y-1. It flew for the first time on April 9, 1953. It may not have sported pontoons, but it made quite a show on water skis (which were called "hydroskis"). It was an experimental (big surprise) fighter jet, and had delta wings. Although the XF2Y-1 didn't break the sound barrier, its developmental successor, the YF2Y-1 did exceed Mach 1 in a dive on August 3rd, 1954. The XF version had two Westinghouse J34 turbojet engines, each with 3400 lbf thrust, and the YF had two J46 engines with 6000 lbf thrust each. The Navy ordered three more but the seaplane fighter concept was dropped soon afterward.

Convair XF2Y-1 Sea Dart

Answer to Question #9: True, although it was actually a glider. (And, it may be a stretch if the only passenger was the pilot.) Made entirely out of paper, glue, and some masking tape (probably a good deal of it), it was an aeronautical teaching aid at Ohio State, and flew on three separate occasions during the month of August, in 1970. It was towed by a car. It didn't have wheels, of course. (How could it have a strong enough axle, after all, if it was made of only paper?) It slid over a grass field on the waxed paper bottom surface of its fuselage (actually cardboard and not strictly speaking, "original paper"). It had a maximum airspeed of 60 miles per hour.

Answer to Question #10: The right answer is choice C. The Bel Geddes airliner design was real, but it was never built. (There were other designs afoot, as well. The Germans made plans for a huge bomber, the Mannesmann Poll, a ten-engine tri-plane. (Another design by Dr. Edmund Rumpler also had 10 engines.)

Dr. Claudius Dornier's Do.X arriving over New York City, 1931,
and artist's drawing of Norman Bel Geddes' Airliner #4

Answer to Question #11: Would you believe E? The booby prize goes to an Italian, Giovanni Battista Caproni, who did design some other very well-known and successful airplanes, but who also had his share of losers. The Caproni Ca 60 Transaereo would be a good contender for the biggest boner. This monstrosity, circa 1920, had *three sets* of *triplane wings* bolted to a 100-passenger flying boat hull, and was powered by eight 400-hp American Liberty engines in pusher and tractor sets. The 9000 square foot wings were equipped with ailerons, and the rear set were elevators. It flew just once, a short straight-ahead flight, upon landing it was damaged and thankfully, it sank.

photographs of the nine-winged Caproni Ca 60 Transaereo, ca. 1920, and the Johns Multiplane, which had seven

Answer to Question #12: Almost true. Well, depending on what you expect by the term "approaching" and also "some aircraft". The answer might arguably be: true. The best (and only) example of that would be one Felix Baumgartner, Austrian athelete and somewhat notorious BASE jumper. He has a specially designed "wing suit" with a carbon fiber delta wing affixed to his back, which might evoke images of some cartoon superhero, with which he successfully *glided across the English Channel*. The predicted glide ratio is 6:1. (This is about what some jet fighters have, actually.) The best glide speed is about 135 knots. Without dealing with any details on just how he would flare for a landing at that speed, he began his glide after exiting an aircraft at a starting altitude of about 27,000 feet. After an initial free-fall straight down to pick up speed, he then completed a gliding flight to the other side, deploying his parachute at about 1800 feet. (see www.felixbaumgartner.com.)

Answer to Question #13: Well, it wasn't Nelson Rockefeller; the former vice president of the United States under President Gerald Ford wasn't the first member of the Mile-High Club. (Also, the year 1979 *does* seem to be a bit late for such a hallmark of audacity, doesn't it?) He did do something somewhat scandalous, though: he died (at age 71) while somewhat aloft in his Fifth Avenue apartment in Manhattan (in the act of adultery with a woman 45 years his junior). Rockefeller's wife, Happy, wasn't too happy when she heard the news; she had him promptly cremated, 18 hours later. It wasn't Howard Hughes, though he probably *did* join the club. After all, he was a glamorous pilot, rich and famous, an engineer in his own right, and probably slept with far too many Hollywood starlets. Rudolph Valentino was a famous silent film era heartthrob, but didn't have much in common with airplanes. The correct answer was the earliest one, too, back in 1916. It was Lawrence Sperry (choice C), the fearless and adventuresome son of Elmer Sperry of gyroscope fame. As handsome as any leading man, *and* an engineering wizard, he invented the turn-and-bank indicator as well as the autopilot. While flying over Babylon, New York with Manhattan socialite Mrs. Waldo Polk, who was as enamored of flying as he was, he put his automatic pilot to the test. Unfortunately, they didn't quite make it and as a result of something going amiss with the stabilizing mechanism, duck hunters soon plucked two start naked swimmers out of the Great South Bay. (If you'd ever like to read about it, here's the book.) And by the way, one doesn't actually *need* to be a mile high to join the Club (or so they say).

photograph of the front jacket of "Gyro! The Life and Times of Lawrence Sperry" by William Wyatt Davenport, published by Charles Scribner's Sons, 1978.

Answer to Question #14: Blind people learn to do amazing things. Just the idea of being able to rapidly read little embossed bumps on a page with one's fingertips is truly amazing (although the popular use of Braille is falling off a bit). It is a known fact that the remaining senses become keener, in compensation. Miles Hilton-Barber is the Project Manager for the Royal National Institute for the Blind. For him, flying blind has literal and not just figurative meaning. He became the first blind person to fly across the English Channel (in a microlight aircraft) on August 24, 2003. Hilton-Barber was among about 80 pilots to take off from Headcorn Airfield, Kent, in possibly the largest peacetime aerial Channel crossing. He did fly with a sighted pilot for safety (and not too surprisingly, legal) reasons.

To paraphrase the reaction of a correspondent in California (who actually *is* blind), she said (wisely, I think) that flying blind is okay when it's done with PAR and with a controller, but. . .landing? She thought there might be an issue there, because of the need for depth perception and distance calculation. (As she pointed out to me, even sighted pilots hit each other.) Of course, wisely, as noted, Miles did fly with a sighted pilot, so we can't know exactly how much of the landing was "his". That said, Miles is still quite motivated. (And why not? He's a motivational speaker!) In April of 1999, Miles completed the "Toughest Foot-Race in the World", the Marathon Des Sables, a 150 mile ultra-marathon through the 120 degree heat of the Sahara Desert. In 2001, he climbed to a height of 17,500 feet in the Himalayas. He then successfully conquered Kilimanjaro, the highest mountain in Africa, at 19,340 feet. He was also the first blind person to manhaul a sledge 400 kilometres across Antarctica, has done white water rafting in Zambesi, tandem cycling marathons, SCUBA diving, and among other things, hot-air ballooning.

Another of his goals is to fly the 12,600 miles to Sydney, Australia, in the same aircraft. It's his way of honoring 100 years of powered flight and raising money for Royal National Institute of the Blind. "Apart from it being a great adventure for me, fulfilling one of my childhood dreams, it's also hopefully helping a lot of other people in Britain to have a better quality of life," he told BBC news. All that blissful and altruistic warm fuzzy stuff aside however, it is absolutely and forever inherently impossible for even a person blind since birth to overcome the physical limitations of human balance and proprioception of the semicircular canals in the inner ear. This brave soul, bless his heart, couldn't have done it alone. The answer is: false.

Photograph of Briton Miles Hilton-Barber in front of G-DCMI,
his microlight aircraft

Answer to Question #15: Choice (D) was pure baloney. But choice (C) wasn't. Can you imagine spending *sixty-four straight days* in continual flight—that's more than two months, folks—inside a 172? (Think of how much <u>time</u> you'd build, though!) The pilots were John Cook and Robert Timm. Note that if the question had asked about the longest *non-refueled* airplane flight, of course the answer would be choice (B), the famous Voyager (which incidentally had an empty weight of about 1021 Kg, less than *one-fourth* of its gross weight, with fuel, of 4397 Kg). If you considered any category of non-refueled aircraft flight, it was choice (A), which was the first around-the-world balloon flight: that of the Breitling Orbiter 3 between March 3 and March 21, 1999. John Cook and Robert Timm incidentally flew the equivalent of six times around the world during their flight. Their approach to refueling was fairly simple; while flying just above the ground, a fuel truck followed the plane closely as fuel was pumped aboard. The pilots lifted baskets of food into the plane by rope. As for the other domestic accoutrements, let's just say they were utilitarian.

Chapter 10: Curiosities, Factoids, and Helpful Hints

Here are a gaggle of eccentrics. Some are indeed trivial, and others are the sorts of things that could very easily prove to be more than just Fun Facts To Know And Tell. Sure, you may find a fellow trivia buff at your next aviation gathering, and that's where some of these might serve: like the highest clouds, or what General George Armstrong Custer had to do with airplanes, or the speed of the "rabbit" from a sequenced flashing light approach system. Others, like what clear air turbulence actually *looks* like, how to combat engine corrosion, the first thing you should do when experiencing induction ice, or the subject of fire extinguishers, are a bit more than just Hints From Heloise.

Subject: Close To Cloud Nine

Question 1: What is the highest cloud type?

A) cirrus
B) cumulonimbus
C) meteoric vapor trails
D) noctilucent clouds

Subject: A Bit Of Inorganic Chemistry, And the Lightest Metal

Question 2: Lithium is the lightest metal, much lighter even than magnesium. It's extremely light (about half the weight of water), with an atomic number of just <u>three</u>—the next lightest element being helium! So why don't they make aircraft parts out of it? (More than one may apply.)

A) It isn't strong enough. With a tensile strength approaching halvah, or maybe torrone, anything made of it would collapse on itself under its own weight. (Having a Mohs hardness of 0.6, it can be cut with a knife.)
B) It's too expensive. Aside from being the lightest solid element (and paradoxically having the highest specific heat of any solid), it's just too dear even to make airplanes out of (though it *is* alloyed with other metals).
C) Being an alkali metal, it's quite reactive with water. As soon as you flew into rain, your entire airplane would outdo the Wicked Witch of the West, and actually explode.
D) All three

Subject: Here's One That's Worth A Few Shekels:

Question 3: If your airport happens to be one of those where the ol' fuel truck comes over to your hangar or tiedown spot, rather than having a pump that one must taxi up to, how might you save a few cents each time you get a top-off?

A) Be sure to tell them NOT to top it off. (This applies also if you've got a bolted-down gas pump, and you're pumping it yourself.) With some line crew personnel, although most are very professional, sometimes they're not too dainty with that hose and you can get an "oops" runover, all over your nicely waxed airplane, and some not-so-nice, OSHA-unfriendly, impromptu weed killer, all over the ground. But mainly, gasoline (car or plane) expands with warming, and unless there's room for expansion, guess where it's going to go if it warms up? Right out the overflow vent! (unless of course you go flying right away)

B) Right after they're done pumping, the meter will read X-point-Y gallons. The line guy writes it up, and hands you the receipt. But then he rolls the hose back up, which causes some flattening on the spool, and it's not unusual for about a tenth of a gallon to get squeezed right back into the truck. That's over half a buck (these days) that you might have paid for, but didn't get. The meter, if it's working, would actually run backwards and show this, by the way.

C) If it's spring or summer, get your tigers in the morning, when the gas isn't as warmed up from the sun's heat--and is more dense. The meter doesn't measure molar volume, y'know--it just counts gallons. This does not apply so much to the fixed pump scenario if they've still got those underground tanks. But those are going by the wayside. And see "A"!

D) A, B, and C

Subject: Wascally Wabbits:

Question 4: At major airports, you will often see varied types of approach lights. Most burn steadily, some flash. On approach to some larger runways, you might see a rapidly moving sequence of lights, similar in effect to the optical illusion of an old fashioned movie marquee. The common term for this is the "rabbit". And those lights move ***fast***! Just how fast *are* those sequenced flashing lights moving, anyway?

A) about 400 knots

B) faster than ANY airplane will ever travel, down this low! The speed is about 1600 knots.

C) Faster still: anywhere between 1.4 and 2.4 nautical miles per SECOND!

D) Warp three

Subject: IAP Navigation via Euclid

Question 5: What's the connection between a non-RMI DME arc procedure and an 18-sided polygon? (okay, pilots only)

A) None. If you don't have an RMI---and most of us don't---one way to fly a DME arc procedure is to chase the DME readout by quickly altering course left or right as need be, whenever the DME readout changes by 0.1 mile. (It requires quite a lot of REMs, though!)

B) Nonsense. That instills about as much serenity as working on a bomb squad---too much work! Instead, a good approximation is to fly successive arcs of a polygon inside the arc's circle. If you can imagine inscribing an 18-sided regular polygon inside a circle, how many degrees does each chord line subtend? Lessee...360 divided by 18: TWENTY! OK, so the sum of the interior angles of each of those 18 isosceles triangles being 180 means that each of those equal "base" angles is: 80 degrees, right? Then it's supplementary neighbors (separated by and on the other side of the chord line) is exactly 100 degrees! So, you fly up to the next radial on the plan view, and simply add (if clockwise) or subtract (if counterclockwise) 100 degrees to get the next 20 degree chord line to fly! In all DME arcs, that puts you well within the allowed tolerances. (For those math nerds out there, "well within the allowed tolerances" is true because the worst deviation of a 20 degree chord line from its corresponding arc is only 1.5 percent of the radius. If you want the exact answer, just take the cosine of 10 degrees.) Anyway, this method involves MUCH less headwork!

C) TERPS criteria require that the lead radials be exact multiples of 20 degrees apart.

D) None. There is no requirement for lead radials (or any other) to be any set number of degrees apart.

Subject: "Away, Away With Rum, By Gum!"

Question 6: What is the significance of a "rhumb line"?

A) It's a great tavern in Newport, Rhode Island. . .also a nice lodge up in Kennebunkport! But seriously, a rhumb line was the straight line that inebriated sailors (or airmen) were challenged to walk after they'd had a few.

B) A rhumb line between two places does not change true course. A rhumb line makes the same angle with all meridians it crosses and appears as a straight line on a sectional (while a great circle appears curved). On a gnomonic projection, it's the reverse, incidentally.

C) Also known as a loxodrome, or spherical helix, it intersects all meridians at a constant angle. On a Mercator projection, it appears as a straight line, but strictly speaking, NOT on a sectional. That's because a sectional (and even a TAC chart) is a Lambert conformal conic projection, which is a cone whose apex is over a pole and which slices the earth at one (or two) chosen latitudes. The result is that the meridians on "our" charts (as well as in reality) are not really exactly parallel!

D) The "rhumb line" (a bit of a play on words) was the route flown by pilots transporting alcoholic beverages during Prohibition--particularly in the Northeast. It actually wasn't anything close to a straight line.

Subject: You Take the High Road, I'll Take the Low Road...

Question 7: What *really* happens to the air when it is hit by your airplane's oncoming wing? (We're talking here just about the air that is in the immediate vicinity of the wing's leading edge, as well as within a short distance above and below it.

A) Any given parcel of air that gets "split" by your wing will of course take two paths. Assuming that you don't have a symmetrical airfoil and also that the wing is developing some lift, the air that goes over the wing must travel faster than that which goes beneath the flatter surface underneath (since it must travel a greater distance) in order to meet up with it again at the trailing edge. The faster moving air, according to Bernoulli's Law, is at a lower pressure, and the suction is what pulls you up into the wild blue.

B) Well, Number One is close, but incorrect on several counts. First, this Bernoulli thing is a Principle, not a Law. Second, there is no such thing as "*suction*" (just like there's no such thing as "cold"---it's just an absence of heat!) But thirdly, and most annoyingly, the air at the top never quite meets up with the air at the bottom because it travels faster, yes, but not fast enough to meet up with the air below it. It lags just a little behind, but due to the elasticity of air it is stretched a bit and lowers the pressure enough to generate lift from below.

C) Number two, but vorticity from the wing tips pulls the "upside" parcel outboard of the lower parcel, so they never actually meet up, anyway.

D) Number one is wrong: they don't meet up. Choice (B) is right, at least about "principles" and "suction". But actually, the air traveling over the wing can accelerate to almost twice the "lower" parcel's speed (at lower airspeeds and high cambers, at least) and almost always arrives at the trailing edge ahead of the lower one (as long as lift is being developed).

Subject: Interception

Question 8: Most of us know what to expect if we were intercepted during the daytime, but what would you see at night?

A) The rules of engagement are unchanged.

B) The intercepting aircraft takes up a position to the right and in front (as opposed to the left).

C) If night VMC, the same, but including the flashing of navigational lights at irregular intervals.
D) flashing of the landing lights from the rear by the intercepting aircraft, then overtaking from above and assuming escort position off the right wing

Subject: The World's Biggest Propeller

Question 9: What is the world's largest propeller? (This does not include helicopter rotor blades.)

A) The F4U Corsair had the biggest propeller ever fitted to a fighter, over 13 feet in diameter.
B) The R.M.S. Titanic's port and starboard propellers were each 23 feet across.
C) High altitude propeller-driven aircraft often have larger, slower-moving propellers. The largest of these approach 20 feet in diameter.
D) There are actually three bladed propellers in excess of 200 feet diameter, tip-to-tip.
E) A Linke-Hofmann R II built in Breslau, Germany used a 22 ft 7.5 in diameter Garuda propeller. It flew in 1919 and was driven by a 260hp engine at 545rpm.

Subject: Rainbows

Question 10: Rainbows are the occasional wonder when the sun pops back out again just after a rain. We usually see just one bow, but somewhat more rarely, it is possible to see

A) trick question. only one at a time...
B) two; the second "supernumary" bow being outside the first
C) three
D) four!

Subject: X-ray Glasses

Question 11: If you could see clear air turbulence, what would it look like?

A) a rotating vertical column of air
B) a dome of rotating air, with the most extreme movements at the top
C) a wave.
D) a "streak" offshoot, branching down from an overhead jet stream

Subject: "So Many Things I Would Have Done, But Clouds Got In My Way"

Question 12: Which of these locations has the greatest possible amount of sunshine?

A) the North Pole
B) the Sahara Desert
C) the Antarctic
D) the U.S. Southwest

Subject: Yessir, We Go 'WAY Back

Question 13: When Chuck Yeager first broke the sound barrier on October 4, 1947, he had another pilot on his wing (who at the time was equally as unheard-of). It was

A) Jackie Cochran
B) Bob Hoover
C) Duane Cole
D) Art Scholl

Subject: Getting Rusty

Question 14: What is the best way to combat rust in an aircraft engine?

A) Fly every chance you get.
B) During any period of inactivity longer than a week, run the engine for a minute or two every three or four days.
C) Top off your tanks to avoid condensation of water in the fuel, which contributes to engine corrosion.
D) Change the oil frequently.
E) A and D

Subject: When Something Snaps:

Question 15: Your best friend in the event of an elevator control failure is

A) the elevator trim
B) carefully shifting passengers and baggage
C) careful power changes
D) all three

Subject: Sunbeams

Question 16: Those ethereal rays of light that shine through gaps or breaks in clouds are called

A) sunbeams
B) crepuscular rays
C) diffraction scatter
D) There actually is no term.

Subject: Colorful Expression

Question 17: Which color of highlighting marker should you use to mark your routes on flight charts: yellow, green, blue, or pink?

A) It doesn't matter. Vary the colors for variety (especially if you have used the same chart for several prior flights).
B) During the day, it's irrelevant. At night however, pink actually works best.
C) Yellow is actually more noticeable than any other commonly used color for highlighting markers (green, blue, or pink). It also shows up better at night.
D) at night, blue; during the day, yellow

Subject: Glory Be!

Question 18: A "glory" is the common name for what visual phenomenon?

A) a rare triple rainbow, seen only in very moist tropical air.
B) the rare halo seen at one edge of the sun's disc during the start of a total solar eclipse, if there is enough moisture.
C) a circular rainbow surrounding an airplane's shadow often seen by pilots on an undercast
D) crepuscular rays (shafts of light shining through breaks in clouds) formed by suspended atmospheric particles, when seen at daybreak (as opposed to other times of day)

Subject: Handy Hints For When Your Back Is To the Wall

Question 19: For those without fuel injection, what is the wisest thing to do *first*, if the carburetor on your reciprocating engine ices up?

A) Immediately select a full rich mixture and apply full carb heat
B) Immediately trade airspeed for altitude until you reach best glide speed.
C) Immediately lean the mixture.
D) Apply carburetor heat.
E) As soon as practical, surrender your license, because you weren't paying enough attention to the atmospheric conditions in the first place, nor to the dropping rpm or manifold pressure to notice the problem before things *really* got out of hand.

Subject: From the Ministry Of Silly Walks

Question 20: Why would you see two sets of figures on an aircraft's compass correction card, one labeled "DAY" and the other labeled "NIGHT"?

A) You might if you've been a tad lax on the eight hour bottle-to-throttle rule and you're seeing double, because there *is* no such distinction, and you will *never* see a two-column compass correction card.
B) In Great Britain, being much closer to the active nighttime ionospheric F2 layer, which descends to lower levels after sunset, there is significantly greater interference at night (and which is fairly constant and predictable). This requires a second column. This also affects other nations in northern Europe.
C) Oh those zany Brits! Yes those same folks who have brought us Monty Python and John Cleese have decided to use a second column to take into account the differing magnetic fields inside an aircraft when it is operated with and without navigation and panel lights. The compass correction cards in British aircraft contain two sets of figures; day and night corrections are required by Section K of the British Civil Airworthiness Code (BCAR).
D) Although it is technically required under FAR 135.149(d), it is seldom followed.

Subject: Light the Fuse, And Stand Back!

Question 21: What was the world's first rocket-powered airplane?

A) the Northrop MX-324, United States, July 1944
B) the DFS-194, Germany, in August of 1940
C) the Ente, also Germany, June 1928
D) the flying chair of Wan-Hu, China, ca. 1500

Subject: The First In-flight Movie

Question 22: The first regular in-flight motion pictures made their appearance in

A) 1925
B) 1948
C) 1961
D) 1970

Subject: Smoke Signals: Advertising Takes To the Air

Question 23: The first use of airplanes to write messages in the sky was in

A) 1913
B) 1922

C) 1927
D) 1941

Subject: Slightly Higher In Bolivia

Question 24: Airliners usually maintain cabin pressure altitudes somewhere between about 5000 and 8000 feet during cruise flight. But just how does the pressurization system cope with situations where they're landing at an airport with an elevation *higher* than that? (There are several. La Paz, Bolivia, for example, at 13,313 feet above mean sea level.) What does an airliner do, when it's flying along at FL 330, with a normal cabin altitude of say 7000 feet, but it's landing where the terrain itself is significantly higher?

A) During the climb to cruise altitude, cabin pressurization is gradually adjusted, up to and then beyond the usual cabin pressure altitude (but more slowly), until it has reached the expected pressure at the higher elevation destination airport.
B) Cabin pressure altitude remains at no higher than 8000 feet, despite the higher landing elevation, because of operational and engineering standardization in most airliner pressurization systems. In addition, Part 135 regulations would require that at least one of the pilots on "flight deck duty" wear oxygen masks continuously, (and for Part 121, that's *all* crewmembers) whenever the cabin pressure altitude exceeded 12,000 feet. Shortly before the doors are opened at the gate, outflow valves allow a fairly rapid (but not explosive) decompression to ambient pressure.
C) Whenever an airliner lands at *any* such high elevation airport, the cabin is actually "depressurized" to that pressure altitude (from the usual cabin pressure altitude), but rather than doing this at the arrival gate at the last minute, it is usually done during the approach to landing. (The oxygen-related regulation mentioned in the previous selection may still apply.) In the case of extreme examples such as La Paz, the cabin pressure altitude can also be raised to 13,300 feet while enroute, at whatever flight level is chosen or assigned. Once the cabin is depressurized to that altitude, the crew maintains that cabin altitude. This is so that upon landing, cabin altitude and actual elevation are the same. The outflow valves automatically open upon touchdown to eliminate any possible residual pressure that might prevent the doors from opening (although in some installations, there are automatic pressure controllers which gradually decrease cabin pressure altitude during the approach, rather than the more rapid drop via outflow valves).
D) Most airliners serving such high (and usually remote) airports do not actually fly at higher flight levels that often, and they simply never pressurize the cabin at all. Also, in most of the countries having such high elevation runways, our FAA rules don't apply, and the crew on such flights are themselves acclimatized to the thin air.

Subject: Little Bighorn, A Century Later

Question 25: For what clever but ill-fated design was General George Armstrong Custer's great grand-nephew best known?

A) a reciprocating aircraft engine that ran on methane, and whose emissions consisted of *only* carbon dioxide and water
B) for founding the only public use airport located within the borders of both a National battlefield monument *and* an active Indian reservation, (in Hardin, Montana), but mostly for crashing his airplane during a landing attempt there exactly one hundred years to the day after the historic battle of Little Bighorn on June 25, 1976. (Actually, the battle took place during both June 25 and 26, 1876, but you can appreciate the irony.)
C) a unique STOL airplane wing design
D) for designing and flying the first rotary-wing aircraft whose main rotor (having four blades) could be spun down in flight and function as a pair of forward and rearward-swept wings

Subject: Too Much Of A Good Thing

Question 26: True or False: Oxygen can actually be harmful to breathe.

Subject: Airfoils

Question 27: You'll often see the letters "NACA" followed by a string of four or five digits, in reference to a particular airfoil. What is the meaning of the numerical designators referring to commonly used airfoils?

A) Their various digits indicate camber and thickness.
B) They are arbitrary numbers assigned in sequence.
C) They represent when they were certified; they're Julian dates.
D) It means such a reference is obsolete; NACA became NASA in 1958.

Subject: Fire Drill

Question 28: If a fire were to somehow erupt in your aircraft, which of the following correctly states how your fire extinguisher might save the day?

A) If the fire were electrical in origin, the discharged water would safely cool the insulating material from associated wiring below its melting temperature such that the likelihood of re-ignition is greatly reduced.
B) If the fire originated from combustible materials, the discharged gas only needs to lower the oxygen content of the air by about a third, rather than "smothering" it by totally displacing the oxygen in the air around it.
C) For combustible metals, carbon dioxide fire extinguishers are effective agents, although they do leave quite a mess afterwards.

D) For flammable liquids, the fire extinguisher does just that by fully displacing the burning gasses that have been evaporated by the heat from the fire with inert gas which does not support further combustion, as well as diluting any remaining volatile liquid with enough of a non-flammable component to prevent re-ignition.

Subject: This Just In...

Question 29: On approach to landing at one of several airports in the Detroit metropolitan area, you hear ATIS Information Zulu. This means

A) It's very late in the day, or the weather has been very unstable, or both. Twenty-six letters is a lot of letters to go through (nominally, at one per hour).
B) It *might* be late in the day, but that doesn't mean the weather's been changing; not every letter of the alphabet is used at each of them.
C) The weather's probably not great, but at most, 18 letters can be used in any 24-hour period, regardless of conditions.
D) nothing. The ATIS cycle period is independent of a 24-hour day.

Subject: Antique Instruments

Question 30: The instrument rating is how old?

A) almost as old as powered flight: the first instrument rating was issued in 1918.
B) not quite, but there are precious few pilots flying today who had an instrument ticket when they first came out...three quarters of a century ago.
C) about half a century: the first instrument rating was issued in 1953
D) forty-two

Subject: The *Real* First Female Pilot

Question 31: The absolutely and undeniably first woman pilot in the whole world was

A) Jackie Cochran, 1938
B) Harriet Quimby, 1911
C) The Baroness Raymonde de la Roche, 1910
D) Aida d'Acosta, 1903
E) Jeanne Labrosse, 1798

Subject: Big Words

Question 32: Avgas is colored by the use of a dye to indicate its octane and lead level. True or false: The higher the octane, the greater the number of letters in the name of the color used.

Subject: All Shook Up

Question 33: As you are about to flare for landing, just when you reduce propeller rpm, you feel a slight shudder go through the airplane, the cockpit, and you. Is there anything wrong?

A) No (although it depends upon how you define normality). Many piston aircraft engines have some degree of inequity in the fuel distribution of the induction manifold, which induces power differences among the engine cylinders.
B) Yes, although the malady isn't catastrophic. Although almost no piston engine crankshaft is balanced to absolute perfection, after a decade or so, about half of the crankshafts will need at least some dynamic balancing. (Again though, smoothness is subjective.)
C) Yes. This can happen most often when reducing power below 1800 rpm, while still above idle speeds (between 1200 and 1800 rpm). While not serious, this is most often because of a propeller imbalance. The majority (over three quarters, by one estimate) of general aviation propellers are, at least to some measurable degree, out of balance. What you may feel occurs when a slight vibration of an out-of-balance prop is picked up by the inherent harmonic "dampening" frequency of the engine mounts themselves.
D) No. You're feeling the interaction of the propeller slipstream and the increased airflow above the wing when entering ground effect.

Subject: Drafty

Question 34: True or false: When the outside temperature is low, closing the cowl flaps prior to, during, or just at engine start (and keeping them closed prior to takeoff) is a smart way to help keep some extra heat in the engine compartment to help warm up the engine. (But you must remember to open them prior to takeoff!)

Curiosities, Factoids, and Helpful Hints: Answers

Answer to Question #1: The right one is choice D; noctilucent clouds--they are typically found at altitudes of around 50 miles. They are usually only visible in high northern latitudes at twilight. (The next highest up are nacreous clouds.) There was a great article about it in an old Scientific American, if you're intrigued enough (it's *really* old; like June 1963).

Answer to Question #2: The answer is A...and, well, B. Sodium is the alkali metal that “melts” in water; potassium actually burns; but pure lithium is also reactive. All three give off hydrogen when in contact with water.

Answer to Question #3: The answer (which you probably guessed from the impassioned verbosity) is D--all three.

Answer to Question #4: The answer is C. They fire based on a standard 60 Hz cycle. Anywhere from 5 to 21 lights are arranged in a row over a 2400 to (sometimes) 3000 foot distance, firing in sequence from the outer-most to the inner-most fixture. In the ALSF mode, they fire every at every cycle of the 60 Hz line--in the SSALR mode, every other flasher fires on every other cycle. The speed ranges between about 8500 to 9600 fps in the ALSF mode to 11,250 to 14,400 fps in SSALR mode. Though in general, the sequence occurs twice a second, there's a "rest time" for old Bugs in there, and the actual speed of the rabbit is much faster. Note: This is for the USA only. Transport Canada systems differ from those in the US.

Answer to Question #5: The correct one is B. (Geez, what a windbag!)

Answer to Question #6: The answer is C. hic.

Answer to Question #7: It's D...believe it or not!

Answer to Question #8: Although the rules may seem arbitrary, it's C. You would see that F-16 (probably near a stall trying to slow down to your cruise speed) rocking its wings to the left and in front of you, as well as flashing its nav lights, followed by a gentle turn (usually to the left) to the desired heading. In acknowledgement, you had better rock your wings also, and follow. At night, you would flash your nav lights as well in response.

Answer to Question #9: D for this one: The gigantic airplane propellers slowly spinning (at about 12 to 19 rpm) atop tall, sequoia-sized towers in a few windier parts of the world are actually functioning as windmills, not fans. They are there to generate electricity, and they do so in Megawatt multiples. Their propellers are actually just the same shape as airplane propellers, although they are usually fiberglass composite, and each blade weighs five tons! They do shut down if the wind reaches about 55 mph.

Answer to Question #10: Theoretically, it would be D. Practically speaking, it's B. Rainbows are formed by the refraction of light inside water droplets, and are found at precise angles. The primary bow forms about 23 degrees from the anti-solar point, which is the point in the sky or on the ground which is exactly opposite from the sun (as seen by the observer). The secondary bow is much fainter, and is at 35 degrees from the anti-solar point. The third order bow is much fainter still, and you would have to be inside the scattering cloud to see it, because its location is 120 degrees from the anti-solar point. The fourth order bow is too faint for the human eye to see but would be at 160 degrees.

Answer to Question #11: It's choice C: So-called Kelvin-Helmholtz waves occur in shear layers near thunderstorms and jet streams. When moisture exists at the boundary (a rare occurrence), the moisture acts as a tracer, and cloud forms (sometimes called "billow clouds") take on exactly the same appearance as a series of cresting ocean waves. Known as Kelvin-Helmholtz instability, it results from velocity shears between two media. (They don't even have to be of different densities.) The most familiar example of this is wind blowing over calm water. Here are actual photographs of this phenomena, the first taken in January 1971 West of Fort Collins, Colorado, and another equally striking one below, taken over Laramie, Wyoming (from the Advisory Circular AC-0057, "Hazardous Mountain Winds and Their Visual Indicators")

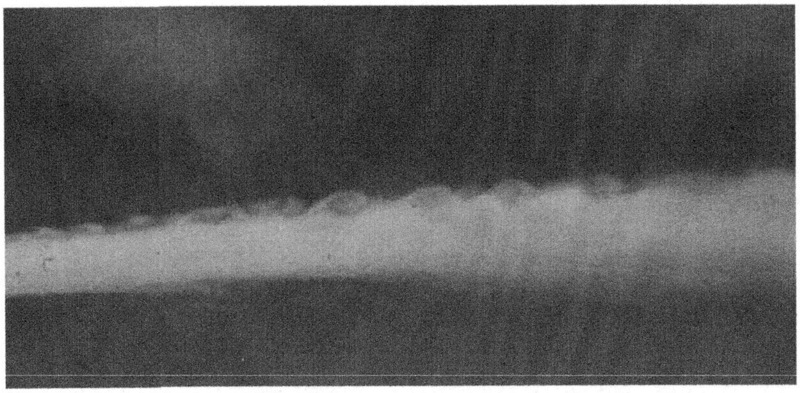

reprinted from NOAA Photo Library, Historic NWS Collection

from Advisory Circular AC-0057, "Hazardous Mountain Winds and Their Visual Indicators"

Answer to Question #12: The right one is choice B. The Sahara leads the world in sunshine, with more than 4300 hours per year. That's 98% of the time. Note: 24 hours/day x 365.25 days/year = 8766 hours/year, but you must consider that about half of that is night time. At the approximate latitude of the Sahara, 25 degrees North, there are about 4383 hours of daylight per year. Incidentally the hottest temperature ever recorded, although it was not official, was about 136 degrees Fahrenheit (in the shade) on September 13, 1922 in Al' Aziziyah, Libya, which is near Tripoli and the Mediterranean Sea, and is actually further North at about a latitude of 33 degrees N. (I hear that hangar space is really cheap there, too.)

Answer to Question #13: The right choice here is B. Hoover was also Yeager's back-up pilot.

Answer to Question #14: It's E. Flying as often as possible is one part of it. As your engine sits idly by, warm moist air in the crankcase cools down at night and the moisture in the air can form water droplets on internal engine parts. Over time this results in accumulations of water in the oil, which can cause rusting of cylinder barrels, connecting rods and other internal parts. Whenever an engine reaches its normal operating temperature, the water boils off. But just running it up for a few minutes and then shutting it down is likely to increase the amount of moisture in the crankcase. Flying for at least 20 minutes allows accumulated moisture to boil off. The second way is by regular oil changes to flush out contaminants. A rule of thumb is to change the oil every 25 hours if your engine doesn't have an oil filter and every 50 hours if it does. If you fly infrequently, changing the oil at least every four months is a good idea.

Answer to Question #15: The correct answer is D. (And also ATC—declare an emergency!) Also, just remember that if the elevator is "floating", the trim wheel will work just like the yoke does, but if the elevator is jammed, then the airplane will respond opposite to trim wheel movement.

Answer to Question #16: B. And they're visible because the sunlight scatters around dust and other particles suspended in the atmosphere.

Answer to Question #17: D is the one. Don't use yellow or pink highlighters at night; they will not show up well with red cockpit lighting. (If you use white light, you *may* see your highlighting, but of course, you'll lose your night vision.) Use blue instead; under red lighting, it shows up as purple. (And yellow does catch the eye better during the day.)

Answer to Question #18: Try C. Originally known by those ascending certain peaks in the Harz Mountains of Germany as the "Specter of the Brocken", it was first considered to be quite mysterious. A climber would ascend above the tops of clouds into the sunlight, turn around, and see a giant on the clouds below with multi-colored rings around his head. The giant was just the climber's shadow, and the rings were circular rainbows, formed from the scattering and refraction of light inside small droplets of water, as is any rainbow, centered at the climbers "antisolar" point (which is of course where his shadow was). Today of course, pilots are the climbers, and their airplane's shadows are the giants (though if the airplane is high enough above the clouds, only the "glory" is seen). In theory, the glory's rings are radially polarized, contrary to a regular rainbow.

In the first picture, the viewer was obviously sitting towards the front of the airliner. The same phenomenon can also occur when looking down on dewy grass towards one's antisolar point, when the sun is low on the horizon, and there is not enough atmospheric moisture to form a proper rainbow. This is called heiligenschein, and it is a bright spot of light around the shadow of the observer's head. The dew droplets on the grass act as lenses, focusing sunlight on the grass leaves, strongly illuminating them, and working again as a lens for the back-scattered light. And on a much vaster scale, shown in the last picture, a similar thing happens in astronomy when the glow of the sun is seen in an opposite direction. The German word for this one is gegenschein ("counter-glow").

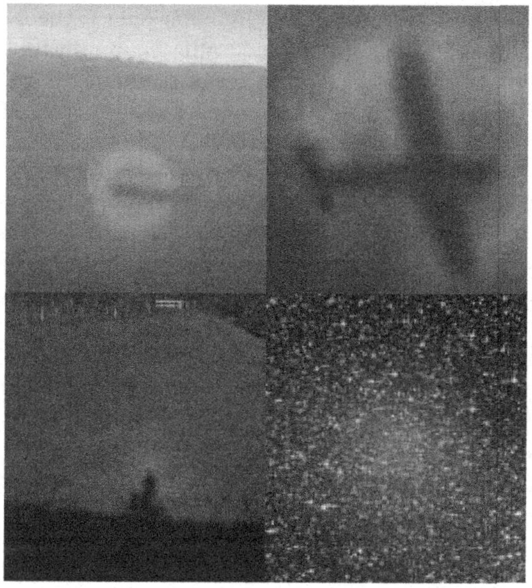

Answer to Question #19: choice C: Ice in the venturi of your carburetor will block some of the incoming air, resulting in a richer than optimal mixture. Yes, apply carburetor heat, but before it takes effect, your mixture will still be too rich, and leaning the mixture to better match the reduced air flow (as well as adding full power) will expedite your being able to help your engine return to normal. The engine will most likely falter momentarily as it ingests water, but afterwards it should return to normal operation (assuming you remember to readjust the mixture for your altitude).

Answer to Question #20: The answer is C. That stuff about the ionosphere regarding compass readings is just so much ozone. (The behavior of the various layers *is* important to radio frequency propagation, but doesn't affect the behavior of a magnetic compass.) The regulation cited from CFR Title 14 Part 135 regarding equipment requirements for commuter and on-demand operations is nothing more than fish bait, because right now, it's still empty (or as they say, "reserved"). Assuming that there *are* such measurable diurnal differences caused by internal lighting, they are not deemed significant by U.S. authorities. This is with possibly good reason: on the various cardinal headings for which separate corrections are given, they each amount to about a degree or so difference. (I guess you could say that they're hardly night and day!)

Answer to Question #21: It was C, the Ente (for "Duck"), which was actually a sailplane, and was the world's first rocket-powered airplane. Powered by two slow-burning Sander rockets and piloted by Friedrich Stamer, it flew about three-quarters of a mile in about one minute on June 11, 1928. (A rocket powered glider was also flown by Fritz von Opel a few months later.) The Northrop MX-324 (choice A) was the first US rocket-powered military aircraft, first flown on July 5, 1944, by Harry Crosby. The -194 (choice B) was designed by Dr. Alexander M. Lippisch of the DFS (Deutsches Forschungsinstitut für Segelflug, or German Research Institute for Gliding Flight) in the late 1930's as a rocket propelled research aircraft. It was the world's first aircraft to be powered by a *liquid-fuel* rocket engine. Early in January of 1939, Lippisch and his project team transferred from DFS to the Messerschmitt company. It was first flown under rocket power at Peenemunde West in August 1940, and made several flights, reaching 342 mph in level flight. The success of the DFS-194 led to development of the Me-163 Komet rocket fighter. Although rockets had been used in China since the West began counting years in a positive direction, nearly all uses were for warfare...or fireworks. However in about the year 1500 according to Chinese legend, an official named Wan-Hu assembled a rocket-powered flying chair. Attached to the chair were two large kites, and fixed to the kites were forty- seven fire-arrow rockets. His success is undocumented.

Wan-Hu with his rocket-powered flying chair

Answer to Question #22: It's A. Various airlines claim to have offered the first in-flight films. The first regularly scheduled in-flight movie made its debut on TWA's Flight 40 from New York to San Francisco flight on July 19, 1961; it featured a sultry Lana Turner with Efram Zimbalist Jr. in "By Love Possessed". (Incidentally, up to the early 1960s, most airlines used 8mm and 16mm films for their in-flight movies.) Before that, as early as 1948, Pan American World Airways advertised their movies "7000 feet above the Atlantic". And back in the 1930's, the in-flight entertainment was sometimes live, such as singers and musicians. But the "We Did It First" award goes to Deutsche Luft Hansa, which began showing single-reel (due to weight considerations) short films on April 6, 1925. They were silent films of course, which was just perfect for the noisy cabins of the day. Choice D, 1970, has no special significance (other than perhaps the year in which the terror-in-the-skies genre movie "Airport" was released...)

Answer to Question #23: That magical combination of smoke and mild aerobatics to write words on the wind which came to be known as skywriting was first added to the realm of advertising in the early years of aviation. It was much earlier than 1941, or even 1927. Trade names and sales slogans were first written in the sky and vividly in memory in 1913, although some references give the year as being 1922. In skywriting, very large letters (typically on the order of a mile in "height" and width, with the actual smoke trail being about 75 feet wide) are "written" (horizontally, and, of course, backwards) by orchestrated movements of specially built airplanes equipped with a smoke-emitting apparatus. Engine heat was usually used to turn specially treated paraffin oil into white smoke, discharged under pressure. Typically performed at altitudes between 10,000 to 17,000 feet, it is most practical in cloudless skies and when there are no more than light winds aloft. (Of course as we know, eventually the wind has its way with those letters, and that 75-foot width becomes smeared out into much wider and then ultimately quite illegible letters.) The usual arrangement is for skywriting to be done over a designated place, and a specified date and time. (Skytyping, the more modern form of skywriting, involves several airplanes, flying parallel and equidistant tracks. The desired message is arranged via a master control panel, and as the planes fly abreast of each other, electronic signals command properly timed emissions of smoke.) But back to the early days...

Some believe that it began in England in 1922 when Major Jack. C. Savage (the cousin of Douglas Savage, a WWI British war ace), originated such early aerial messages. After the First World War, surplus military aircraft were sold at very low prices, and one SE5a was bought by Major Savage, who converted it for skywriting purposes and developed the technique as a form of advertising. Among his first jobs was advertising a newspaper's name, the Daily Mail, over England in May 1922. His first business target in the US was George Washington Hill, the head of the American Tobacco Co. He persuaded Hill to view a rehearsal, and his pilot, Captain Cyril Turner of the Royal Air Force, flew over Times Square, spelling out the words: "HELLO USA" while gawking and spellbound crowds stared up. But Hill was not impressed. "Interesting," he commented, "but it won't sell cigarettes." Yankee ingenuity came to the rescue. The plane went up again, and this time, he wrote "HELLO USA. Call Vanderbilt 7200" (the telephone number of the Vanderbilt Hotel, to which Savage had invited Hill). For three hours the Vanderbilt's switchboard was swamped with 47,000 phone calls. That did the trick. Hill was immediately enthused and signed with the Skywriting Corporation of America for a million dollars worth of skywriting. Note: Some references attribute this stroke of marketing insight to Captain Allen J. Cameron, an American pilot. But regardless, *neither* of them were the first. On July 19, 1913, in the skies over Seattle, Washington, Milton J. Bryant began a new form of advertising: skywriting. The right answer is choice A.

Answer to Question #24: Enough blabbing already--the correct answer is choice C.

Answer to Question #25: It's C (for Custer) The Custer Channelwing ("CCW") was an STOL design wherein each of two engines was placed above and within a half-cylinder shaped airfoil, each of which behaved like the lower half of a venturi. Because the speed of the air drawn through these channels was much greater than the velocity of the airplane through the air, a vastly increased lift was possible at a very low airspeed. (After all, Bernoulli's Law states that the dynamic pressure of a moving fluid is affected proportional to the *square* of its velocity!) Although the CCW still needed some small distance (under 200 feet, with no wind) in which to become airborne, only because some speed was needed for its control surfaces to become effective, it could literally lift off while taxiing. The last and most ambitious version, the CCW-5, had an empty weight of 3200 pounds, an approximate cruise speed of about 180 mph, and even though it was the biggest and heaviest of the CCW designs, it had a landing speed of a little over 20 mph. It was even reportedly able to hover in an 11 mph wind. The Custer Channelwing number one, or CCW-1, first flew on November 12, 1942. The only *possible* technical difficulties were, theoretically, some potential for introduction of asymmetrical loads on the propellers (which was not proven), as they left and entered the lower channeled half of the wing, and an increase in drag from interactions between the channel and propeller which would have limited its top speed to somewhere around 300 knots (350 knots, with minor modifications). Still, it was quite an idea...

Willard Custer was one of the early pioneers in STOL aircraft. He devoted almost 40 years to his design, as well as the company he founded, to make it a reality. Unfortunately, political naïveté, his unwillingness to suffer fools gladly, and his famous lack of diplomacy, which rivaled that of his famous ancestor, caused alienation of potential backers and government contacts. During one particularly frustrating meeting for example, he stood up, pointed a finger at a general, and shouted, "General, you're just a damned liar!" He was offered many millions for his patents, but wouldn't let anyone work with it but himself. Only the CCW-5 at the Mid Atlantic Air Museum in Reading, PA, and CCW-1, until recently located at the Air & Space Museum's Garber Restoration Facility (at Dulles Airport in Virginia, and moved from Suitland MD) still exist.

Willard Custer was not a pilot, and choices B and D were both fictional. There may however have been some association with methane (choice A) and what Mr. Custer thought of the value and contents of his detractors.

The CCW-5, five seconds into its takeoff run. (Reprinted with permission of Custerchannelwing.com)

Answer to Question #26: You might be able to see a curve ball coming from a mile away, but since when could it be dangerous to inhale...when it's life-giving oxygen? With no intentions to mislead, nor to make inferences about the tragic fire inside the Apollo I command module in 1967, it isn't the fact that 100% oxygen is highly supportive of combustion. It also has nothing to do with the dangers of an embolism when one goes SCUBA diving and then goes flying just a few hours later. It has nothing to do with taking a high-pressure oxygen tank up into a low pressure environment. And it isn't because of the dangers of breathing industrial oxygen (which may contain impurities) or medical oxygen (which has water vapor that can subsequently condense and freeze, cutting off the supply). Mother Nature doesn't have a Surgeon General, but the fact is, breathing even pure, 100% aviation oxygen can be harmful to your health. Yes, it's true, as in: toxic symptoms.

And here's how. The catch is, that you'd have to breathe it for quite a long time...like a few dozen hours at a stretch. In controlled experiments, using just 90% oxygen, observed symptoms included nervousness, lowered energy, irregular heart beat, as well as bronchial cough, fever, infection, and vomiting. True, these didn't appear until the second day, and probably no pilot would ever have to be force-fed oxygen for that long a period at high altitude. (The maximum mission duration for even the few elite high-altitude reconnaissance pilots has been somewhere around 14 hours, which would still not be long enough.) Interestingly, switching to pure oxygen after an incident of decompression can actually aggravate the symptoms of hypoxia, which isn't exactly intuitive. (The best remedy is to gradually increase the oxygen supply.) This phenomenon is actually known as oxygen poisoning.

Answer to Question #27: A wing is designed to produce lift and minimize drag, and its shape is the most important feature in fulfilling this role. Almost any cross sectional shape (meaning the airfoil) will produce lift. Many present-day airfoils are wing cross sections described by the early NACA organization (the National Advisory Committee for Aeronautics, chartered in 1915, operational from 1917-1958, and from which the National Aeronautics and Space Act of 1958 created NASA). When NACA began its investigation of airfoils in the 1920s, they started by examining the most successful airfoils of the time. Very quickly they observed significant commonality among popular airfoils, even though they'd at first appeared quite different, and they also realized that simple mathematical equations could represent an airfoil, just by specifying the camber line and thickness distribution. As it was described in NACA's landmark report "The Characteristics of 78 Related Airfoil Sections from Tests in the Variable Density Wind Tunnel" thickness distribution is a single equation, while the camber is usually two joined quadratic equations. As you probably guessed by now, it's choice "A".

The NACA four-digit "foil family" was developed in 1932, and combined three sets of digits: the first digit represents the value of the maximum camber (the amount that the airfoil curves, in percent of the chord width); the second digit is the location of the maximum camber from the leading edge in tenths of the chord width, and the last two digits denote the maximum thickness of the airfoil in percent. They have been tested in many wind tunnels over the years. For example, the NACA 2412 (which you'll see on most Cessna 172s) has a camber of 2% of chord with maximum camber occurring at 40% of chord from the leading edge, and a thickness/chord ratio of 12%. Most airfoils for light aircraft have 2% to 4% camber, thickness ratio of 12% to 15%, and maximum thickness at around 40% of chord. Symmetrical airfoils have only thickness, and no camber. NACA four-digit symmetrical foils are listed as "NACA 00_ _" (and used on rudders, boat keels, etc). NACA five-digit foils were developed in 1935. The primary difference is the use of a different camber line (having more forward camber). In a five-digit airfoil, 1.5 times the first digit is the design lift coefficient in tenths, the second and third digits are one-half the distance from the leading edge to the location of maximum camber in percent of the chord, and the fourth and fifth digits are the thickness in percent of the chord. Modified NACA four and five-digit series designations include additional numbers (such as after a hyphen), indicating other attributes such as roundedness of the leading edge, or design lift coefficient. Later advances involved further six, seven, and eight-digit designators and involved specifying desired pressure distributions over the airfoil and then mathematically deriving the shape producing it; these airfoils were not generated using some sets of analytical expressions as the earlier series. But speaking of earlier, they still say "NACA"; the name stuck, even though there *isn't* any more NACA.

Answer to Question #28: A fire needs three things to get started: fuel, oxygen, and heat. Remove any one, and you stop the fire. Fuels are "classified" into four categories, each denoted by a letter. So-called Class A fuels include wood, paper, cloth, or plastic. Class B fuels are represented by such flammable liquids as gasoline, oil, grease, and solvents. Class C means any "energized" (as in "plugged in") electrical equipment, and Class D fires involve combustible metals such as magnesium, potassium, and sodium. Fire extinguishers are similarly classified. Class A fire extinguishers might use water, which cools the fire by removing heat. Dry chemical or carbon dioxide fire extinguishers can be used for Class A, B, and C fires, and they smother fires by removing (or reducing) the amount of available oxygen. Dry power (such as sodium chloride) extinguishers are used to fight flammable metal fires by covering the burning metal with a crust that forms over the burning metal, smothering the fire.

You might be surprised to learn that fire actually doesn't need the 21 percent of oxygen that is contained in the atmosphere. However, it does need at least 16% of it to consist of oxygen. A standard carbon dioxide cylinder only needs to displace enough air to reduce the percent of oxygen immediately accessible to the flames by about five percent. It doesn't actually have to smother a fire by *totally* displacing all of the air. *That* is why they work so well. So, yes, as far as which "class" of *these* choices is right (A, B, C, or D), the correct answer is B.

Perhaps more important than that 16% stuff though, is to remember that a fire on (or inside of) an airplane is *not* like a fire on or within a surface structure. The type of fires that can occur in a home or business involve normal surface-use "Class A-B-C" fire extinguishers, which have excellent fire fighting capability. Just don't use them in your airplane! Why? Some of the chemical agents melt and flow when in contact with heat. Chemicals such as mono-ammonium phosphate or methyl bromide can be highly corrosive to aluminum alloys and zinc. Once it contacts hot aluminum and flows down into the structural cracks and crevices, it cannot be washed out as other dry chemical agents are. Using an "A-B-C" extinguisher may make it necessary to totally disassemble the aircraft in order to prevent possible destruction of the aircraft by corrosion. (However, "B-C" extinguishers are usually okay.) The best bet, and the safest and least hazardous type (at least as far as you and your airplane are concerned) is to use a halon type of extinguisher.

Answer to Question #29: In metropolitan areas such as Detroit, only prescribed ranges of ATIS identifying letters are used for each airport. That way, when you say you have Information Such-and such, ATC knows from which airport you have received the information. In the case of this city, Detroit Metropolitan (DTW), Detroit Willow Run (YIP), or Detroit Grosse Ile (ONZ) each have separate independent ranges of ATIS letter designators. (The correct choice is B.) Additional airports with "split codes" are in Colorado Springs and Denver, along with other places. Where can you find this rule? FAA Order 7210.3, Paragraph 10-4-1: Facility Operation and Administration: in Chapter 10 (Terminal Operations, Services, and Equipment), Section 4, Services, and sub-paragraph "d" on Automatic Terminal Information Service (ATIS)...it says the following:

> "Specific sequential portions of the alphabet may be assigned between facilities or for an arrival and departure ATIS when confusion could result from using the entire alphabet for each ATIS.
>
> 1. A LOA shall be established between facilities designating the ATIS codes which will be used by each facility.
> 2. A facility directive shall be developed designating the ATIS alphabet codes which will be used by each facility or for an arrival and departure ATIS.
>
> EXAMPLE-Departure ATIS codes could be assigned codes of 'Alfa' through 'Mike' and arrival ATIS codes assigned 'November' through 'Zulu.' The ATIS codes may also be divided between facilities."

Answer to Question #30: Instrument ratings for U.S. civilian pilots were first required in 1936 (August 15, 1936, to be exact). Until then, one could fly merrily along, in the clouds, beholden to no one. It wasn't exactly habit-forming of course, and the Bureau of Air Commerce thereafter required all non-airline pilots to have instrument ratings whenever the visibility was less than a mile. (Pilots could only fly federally licensed aircraft having two-way radios and approved instruments; the airlines and military flew under somewhat different rules.) It's choice B.

Answer to Question #31: Jackie Cochran (choice A) was born in 1906, and won the famous Bendix race from Los Angles to Cleveland in 1938. (She was also the first woman to break the "sound barrier" in 1953.) She was however definitely *not* the world's first female pilot. Harriet Quimby (choice B), the subject of another trivia question, was the first female pilot all right, but only in the United States. She received her license on August 1, 1911. The Baroness Raymonde de la Roche (choice C) received the world's first pilot's license ever awarded to a woman on March 8, 1910. She too has been mentioned in an earlier question. But before there ever *were* pilot licenses, supervised solos, or anything else in the way of certificates and other official credentials, there were airships and balloons. (Who says someone has to have wings in order to be called a pilot?) On July 29, 1903 in fact, in Paris, over four months before the Wright Brothers, Cuban-born New Yorker Aida d'Acosta (choice D) soloed in a dirigible owned by Alberto Santos-Dumont. But crank up the Way Back for this *next* one: On June 4, 1798, one Jeanne Labrosse (choice E) made the first woman's solo balloon flight, in France. For the record, Mme. Elisabeth Thible, a French opera singer, made the first-ever balloon flight by womankind (as a passenger), in 1784 and thus, in a manner of speaking, she became what could conceivably be looked upon as the world's first female pilot. (But that's a bit of a stretch, now isn't it?) The best answer is choice E. Oh, and in case you might be wondering, the first female airplane passenger (for whom there is any record at least) was one Mrs. Hart Berg, the wife of the European business manager for the Wright Brothers. She flew with Wilbur on October 7, 1908.

Answer to Question #32: As we know, the nearly ubiquitous 100/115 avgas has a light blue color. Okay, blue: that's four letters. The next-highest octane level for aviation fuel is 100/130, originally designated as the primary fuel for larger reciprocating single and multi-engine aircraft, and colored green. (This grade is usually also approved as a fall-back fuel for use in most Cessna, Piper, and other smaller general aviation aircraft.) That's also one more letter. Going the other way, one step *below* 100/115 aviation fuel is the erstwhile 80/87 avgas, which was red in color and was used prior to the late 1970s. (Actually, it's still available in some parts of the country.) So we have three colors; we also have three grades, and the color of each fits the pattern. Think that might be just a fortuitous coincidence? Would you still think that if you knew that there also used to be an even higher octane (and high lead) fuel, 115/145, intended for radial engines...and that it was...*purple*? So the answer to this question is true, right? (It sure would be helpful if this "letter rule" were true, to help us better remember, the next time someone asks you *which* grade is *what* color.) Well, it's almost true. The recently released unleaded 82UL, which can be manufactured using the same petroleum stock as automotive gasoline, is also purple. But you're still not likely to see it at an FBO near you, because 82UL is only approved for use in airplanes having a supplemental type certificate or "STC" approving the use of automotive gasoline (with an aviation lean octane rating of 82 or less, or an antiknock index of 87 or less). So...the answer is kinda-sorta true, but technically, false.

Side note: There are several methods of measuring "octane" rating, two of which are known as aviation and supercharge. (Others are research, and motor; the difference is mostly in the engine operating conditions and the measurement method for knocking.) For automotive gasoline, it's the average of research and motor, which explains the "R+M/2" thing on gas pumps. For aviation gasoline, which is highly leaded, the grading comes in the form of two numbers: first from the aviation test (normally aspirated and with a lean mixture), and the second from the supercharge test (with a rich mixture). They're not octane ratings as those used for cars, but "performance" numbers; determining the dependence of the highest permissible power in terms of indicated mean effective pressure on mixture strength and boost for a specific light knocking setting.

Answer to Question #33: The answer is choice (C). Other symptoms include vibrating engine needles, a magnetic compass that continues shuddering and doesn't ever calm down, or even a numbing sensation in one's hands and feet from continual contact with vibrating surfaces, in more extreme cases.

Answer to Question #34: Nope; that is a negative. Unless your Pilot Operating Handbook advises to the contrary, you should always keep the cowl flaps *open* while on the ground (even when the air is freezing cold). Although closing the cowl flaps will speed up engine warming, it could also contribute to uneven engine cooling, which would create *other* problems.

Chapter 11: Regulations In Question

We all play by the rules—most of the time, at least. As a whole, most of us are compulsive law-abiders. On the other hand, most of us, while we're largely optimistic, are also objective and questioning by nature. We're good at integrating information, seeing the Big Picture, and putting details in the proper context. But the typical pilot personality also involves determinants of independence, and the need for excitement, prestige, as well as some control over our environment. For anyone moving around quickly in three dimensions after all, you'd expect no different. So oddly enough, even though we're preoccupied by freedom, we also share somewhat of a fascination with exploring its boundaries. Some limits are academic (for most of us at least, like this first question which deals with Class A airspace); some are trivial (such as the color of the FAA seal on your pilot certificate); and some aren't (for instance, the one about sharing expenses with your passengers). Each one however has something unexpected—sometimes picky, sometimes profound.

Subject: VFR In Class A Airspace?

Question 1: Under what rule in "the FARs" could you fly VFR in Class A airspace? (for non-pilots, that means at 18,000 feet or more above mean sea level, aka flight level 180...up to 60,000 feet)

A) There is no such rule. All operations in Class A airspace must be under Instrument Flight Rules, period.
B) Actually, under FAR 91.159, you can. This rule on VFR cruising altitudes specifically refers to flight above 18,000 feet in section (b), and even flight level 290 in section (c). It spells out which altitudes to use, depending on your magnetic course, using hemispherical rules just as we do below FL 180.
C) True, but another harmless oversight. More realistically you could do this under FAR 91.3, the familiar 'authority of the pilot in command' rule that says you can break any other rule in the book, in an emergency. Say you were cruising along IMC in your T210 at 17,000 feet, with reported tops at FL 190, and you had a total electrical failure. (Don't look to 91.155 for flight visibility and cloud clearance minimums in Class A, though.)
D) Section 91.135 on operations in Class A airspace, paragraph (d) states that you can deviate from any provision of that section with an ATC authorization, although such a request must be submitted in writing at least 4 days before the proposed operation. They may authorize it on a continuing basis, or just for one individual flight.

Subject: Up, Up, And Away!

Question 2: Can you launch a non-tethered balloon (helium, or any other type) without calling the FAA?

A) If it is a toy balloon and it and its payload don't weigh more than four pounds, do your worst. You don't have to tell anybody. If it's over the four pound limit, better look at Part 101 of the FARs. There's a whole bunch of other stuff you'd need to have as well, such as a radar reflector, auxiliary cut-down mechanism (and associated radio receiver and/or computer control circuit), filing NOTAMs, and keeping the FAA informed of planned flight parameters and progress.
B) If it's over six feet in diameter, you have to file a NOTAM. (A six foot helium balloon will lift almost nine pounds, by the way.)
C) There are no rules for balloons under 12 feet in diameter. If you've got a helium balloon that big and you're just fooling around, you have WAY too much free time, and probably more money than brains. (Go find a nice sturdy lawn chair, and have fun.)
D) Any balloon, especially metallized reflective mylar types, will play hob with any nearby radar installation. This is as-yet unlegislated territory, but someone will come along before too long and liven up somebody's afternoon in the tower, I'm sure.

Subject: The Twilight Zone

Question 3: Night, as defined in CFR14, FAR Part 1, is the time between the end of evening civil twilight and morning civil twilight, as published in the American Air Almanac, converted to local time. But what exactly *is* the meaning of twilight?

A) Twilight is when the last chord of the setting sun can no longer be seen below a perfectly flat horizon.
B) Twilight is when the first nightingale sings.
C) Twilight, in this case civil twilight, is when the center of the sun's disc goes one solar diameter below the horizon.
D) Civil twilight means when the center of the sun's disc is six degrees below an ideal horizon.

Subject: N, <u>what</u>? "Vanity Plate" N-Numbers

Question 4: If you want a special N number for your airplane, (even if you don't own one yet or have not finished building it), how do you go about getting one?

A) You have to request it from the manufacturer, if not kit-built, and are limited to certain ranges given in FAR 47.
B) These are only available for amateur-built aircraft, and must however end in two alphabetical symbols.

C) You must pay a special $10 fee, and submit FAA Form 8050-64.
D) There is no form...yet; the 8050-64 is what the FAA sends to you, once it's assigned. (But there is a $10 fee.) You may request up to five possible choices, consisting of between one and five alphanumeric characters.

Subject: More About N-Numbers

Question 5: What dictates the size of the tail number on an experimental aircraft? (More than one may be correct, this time.)

A) Whether or not you take it outside the US or into an ADIZ.
B) How fast it can go.
C) Whether it is classified as "small", "large" or "heavy".
D) It's entirely under the jurisdiction of your local FSDO.

Subject: Painting the Town

Question 6: Which of the following is true regarding FAA rules on obstruction marking and lighting?

A) Any structure, whether or not it is temporary, must be marked and/or lighted if in excess of 300 feet above ground level (including any appurtenances).
B) Alternating bands of aviation orange and white used on narrow structures (as opposed to checkerboard patterns between five and 20 feet square on tanks or buildings) must be odd in number, with the top and bottom bands colored orange, and no more than 100 feet wide.
C) Any failure of a required obstruction light of more than 60 minutes duration should be reported immediately to the nearest Flight Standards District Office.
D) Spheres used to mark wires across canyons or rivers should be no less than 20 inches wide and spaced 300 feet apart.

Subject: Trivial, But Maybe NOT So Trivial

Question 7: A private pilot is allowed to share expenses with passengers. If a flight with two other people on board costs $100 in actual out-of-pocket expenses, what part of this cost must the pilot pay, if any?

A) Not one penny less than one third. (In fact, rounding it down to $33.33 would actually be in violation of the FARs.) It only applies to fuel and oil though.
B) If the flight is in connection with a charitable airlift and meets all seven stipulations as given in FAR 61.113(d); or the flight is a search and rescue flight sanctioned by an approved agency or organization as specified in

61.113(e); or the pilot has at least 200 hours and is an aircraft salesman as specified in 61.113(f), only then can the pilot can pay one third.

C) If "out of pocket" means fuel, oil, airport expenditures, and rental fees, and the flight meets the conditions of choice B, then yes, he or she can ask each of the passengers to pay one third of those expenses. So the pilot can get away with paying just one third, but there are a few conditions in FAR 61.113. They are however not as restrictive as in B; the $100 hamburger is a legitimate mission.

D) Answer C is correct as to the type of expenses, but incorrect as to the proportion. FAR 91.113(c) says that a private pilot may not pay less than the "pro rata" share of these expenses. Aha, but what does "pro rata" mean? It's Latin legalese, of course, and it means "according to the calculated share", or in proportion. What you would expect that to mean is that the pilot could pay one third (okay, $33.34) , and each of the other two passengers could then pay $33.33. Not so. The FAA considers all passengers, however many, as the "other" party. So the answer is... fifty bucks.

Subject: Bending the Rules

Question 8: There is actually a Federal Aviation Regulation (FAR 91.905) that specifies which rules can be waived, in order to operate aircraft in ways not normally permitted under Part 91. These waivers involve many different rules including airspeed and altitude restrictions, and are processed by the local Flight Standards District Office. Aside from helicopters, which are already allowed much less restrictive limits, particularly with regard to minimum safe altitudes and visibility (especially in uncontrolled airspace), what other aircraft category (or categories) may fall under this special rule, and under what circumstances?

A) There are no such permissive changes.
B) Any category may be authorized, under proper circumstances
C) Balloon flights at authorized "aviation events"
D) Only experimental NASA flights

Subject: Who Ya Gonna Call?

Question 9: True or False: There is actually an FAA procedure to report UFOs.

Subject: A Rule, By Any Other Name

Question 10: Officially, FAR stands for

A) Civil Air Regulations
B) Federal Aeronautical Regulations
C) Federal Aviation Regulations
D) Federal Acquisition Regulations

Subject: Tail Numbers And Call Signs

Question 11: True or false: It is possible for an amateur radio operator to reserve his (or her) ham radio call sign for the tail number of his (or her) aircraft.

Subject: I'll Huff, And I'll Puff...

Question 12: Prior to 9/11, how much impact resistance were flight deck doors required to have?

A) resistance to a full running charge by a 350-pound man
B) two blows of approximately 220 foot-pounds of impact force at door, bolt, and hinge points
C) 500 pounds of pushing or pulling force on the panel, or a shear force of 300 pounds at the hinge line
D) A fourth dan jump side-kick aimed at the door lock's deadbolt

Subject: The Fine Print

Question 13: What is Rule 240?

A) the collective set of ASTM and OSHA rules that dictate just how wide (or narrow) airline seats may be, depending on the latest biometric population data
B) the area rule formula discovered by NACA's Dr. Richard Whitcomb, proven during the development of the F-102A, that supersonic flight is best achieved with a "Coke bottle" constriction in the cross-sectional area of the mid-fuselage
C) the little-known rule governing airlines and what compensation they are obligated to provide to ticketed passengers denied a seat for other than "force majeure" events beyond their control (such as weather)
D) the rule dividing the Atomic Energy Commission's Nevada site into numbered areas for the purpose of tracking fallout from nuclear tests...in one of which, starting in about 1954, the Lockheed/CIA "Skunk Works" testing was begun that led to the U-2. (We now refer to it as Area 51.)

Subject: Magna Avia

Question 14: The first aviation law dates back to:

A) 1926
B) 1919
C) 1907
D) 1783

Subject: Status Symbols

Question 15: What is the meaning of the differing colors on the FAA seal in the upper-right corner of various types of pilot certificates?

A) none whatsoever: Some pilots have "blue circles" and some have ones that are black.
B) Supposedly, blue seals are awarded to instrument rated pilots.
C) That, and more: Blue seals are bestowed upon instrument-rated pilots, and "Gold Seal" flight instructors have, well...gold seals!
D) That *would* make sense, wouldn't it? Although there is some truth to this, many non-instrument-rated pilots also have blue certificates.

Subject: Who's On First?

Question 16: After an aircraft experiencing an *emergency*, what four types of aircraft or situations come next, on the pecking order of priority ATC handling?

A) presidential flights, air ambulance flights, *urgent* situations, and search and rescue (SAR) operations
B) any urgency condition, air ambulance flights, presidential flights, and SAR operations
C) civilian air ambulance, SAR, presidential, flight check
D) air ambulance, SAR, presidential, interceptions

Subject: Don't Leave the AME Without It

Question 17: Assuming that he or she has passed the exam, what should a student pilot applicant check, before leaving the office of the Aviation Medical Examiner? More than one may apply.

A) that the AME has given them a Medical Certificate and Student Pilot Certificate (FAA Form 8420-2), and not just an FAA Form 8500-9 (Medical Certificate)
B) that the AME has *not* checked the box marked "Medical Certificate" in block 62 of FAA Form 8500-8 (Application For Airman Medical Certificate or Airman Medical and Student Pilot Certificate).
C) that he or she has signed your FAA Form 8710-2, Student Pilot Certificate, and that you have signed it, as well
D) that the AME has returned your copy of FAA Form 8055-5, Application For Student Pilot Certificate, for your records.

Subject: The "Splitting Hairs" Department

Question 18: As we know, third class medical certificates are valid for 36 calendar months for those of us under 40 years old, and 24 calendar months, if you are 40 years old or more. Okay, so what happens if you are a pilot who is less than three years from being 40? Do you still get three years, even though part of your "medical currency period" runs *after* you've hit the big FOUR-OH?

A) It's good until the last day of the month of your 40th birthday.
B) Your medical is valid for 36 calendar months, regardless.
C) If you are within 24 calendar months of your 40th birthday, your medical certificate is valid until the end of the year during which that 24th month occurs.
D) If you're two years or less from your 40th birthday, you get 24 calendar months. If your 40th birthday is 24 to 36 months off, you get 36 months.

Subject: Are Two "I's" Better Than One?

Question 19: True or False: A private pilot with an airplane, single-engine land rating can always be guaranteed to receive better flight instruction if he gets it from an *instrument* flight instructor.

Subject: Earning Your "Wings"

Question 20: The number of hours of flight instruction needed for a flight review, regardless of whether it is through the Wings program or via a BFR, can be

A) as little as one for a BFR, but always three for Wings
B) for a BFR, two; for the Wings program, two or three
C) at least one for a BFR; and two, three, or *four* for Wings
D) one hour for a BFR, and anywhere from less than an hour to four hours, for the Wings program

Subject: Eat At Joe's

Question 21: Which of the following classes of aircraft are allowed to tow banners?

A) any of them
B) airplanes only
C) airplanes and helicopters only
D) any powered aircraft

Subject: Night Moves

Question 22: You're practicing touch-and-go landings in Class G airspace, at night. Visibility is one and one half statute miles, and you're flying a reasonably tight pattern, with the downwind leg separated from the upwind and final approach legs by a distance somewhere between one half and two-thirds of a mile. The traffic pattern altitude is 1000 feet AGL, and there is an overcast layer 400 feet above the pattern. Which of the following is true?

A) You need at least three miles of flight visibility.
B) At night, takeoffs and landings must be to a full stop.
C) You're too close to the ceiling; you must remain 500 feet below the clouds.
D) none of the above

Subject: Low And Slow

Question 23: You are flying an instrument approach at a stabilized airspeed of 100 knots. However, your aircraft's power-off stall speed (in the landing configuration, at max gross weight) is 65 knots. This would put it into an approach category of 1.3 times that V_{S0} number or about 85 knots (actually, 84.5), which would put your aircraft in approach category A (which is up to 90 knots). You're now flying in approach category B territory though (91 to 120 knots.) Is it required that you fly the approach using the Category B minimums?

A) Yes, it's the law.
B) It wouldn't matter actually, because very few approaches have separate minimums for Category A and Category B (either straight-in or circle-to-land), and the vast majority combine categories A and B.
C) There is no such requirement.
D) There is no regulation specifically mandating this, but it's highly advisable that you do so.

Subject: Chip Off the New Block

Question 24: A pilot who desires to retain instrument currency may use a Personal Computer Aviation Training Device (PCATD) under what conditions?

A) As per Advisory Circular 61-126, paragraph 5.b., under no circumstances can a PCATD be used, except training acquired and properly logged to meet the requirements for instruction for an instrument rating as per Title 14 CFR part 61.65, and for only 10 hours of that requirement. Again, use of a PCATD may *not* be applied towards meeting currency requirements.
B) Only if he or she is already instrument current. Unlike flight accomplished with an instructor or a safety pilot, when using a PCATD, there is no six-month grace period.
C) Only under the supervision of an authorized instrument flight instructor (a CFII, but *not* a "non-instrument" CFI)
D) Only for those PCATD installations within the recently approved "Advanced PCATD" category of aviation training devices. (These have enhancements such as life-sized instruments and an external visual system, as well as improved ergonomics.) This includes the requirements for logged flight experience in Part 61.51(b)(3), the logging of instrument flight time in 61.51(g)(4), and instrument experience as described in 61.57(c)(1).

Subject: Privileged Characters

Question 25: You are on a precision approach into a major metropolitan airport and you have reached the 200-foot ILS decision height without seeing the runway environment, approach lights, or anything else except the inside of a milk bottle. Yet you continue your descent down towards a Category II decision height of 100 feet. Your aircraft is IFR capable, but it is without any of the additional instruments required for Category II authorization specified in Appendix A to Part 91. You are not breaking any rules. How can this be?

A) You are not in the United States.
B) You have earned a Certificate of Waiver from the FAA.
C) You have a synthetic vision system certified under Part 91.205(i)(2)
D) You are a military pilot.
E) You are flying a Precision Approach Radar into a military airport that is approved for joint civilian/military use.

Subject: Noblesse Oblige?

Question 26: True or false: A controller is permitted to issue airborne holds only to IFR traffic.

Subject: Carrying the Torch

Question 27: If an IFR aircraft which precedes you into the same airspace ceases to communicate with ATC for whatever reason, how long will ATC block that airspace from other IFR (or special VFR) traffic—or you?

A) They don't.
B) 15 minutes
C) 30 minutes
D) 45 minutes

Subject: RHIP

Question 28: The amount of time that an airline pilot has logged in his logbook library can be misleading because

A) If a pilot is a captain, he (or she) is allowed to log all time spent on the flight deck as "Pilot In Command" or PIC time, even while the copilot is actually flying the aircraft.
B) The above is true, but the rules are even more generous. The captain does not even have to be on the flight deck. (This is also true in principle for flight instructors in general aviation aircraft, but rather difficult in practice.)
C) Choice (B) is actually true, however, this is allowed only for each leg of the flight for which the captain actually performs the landing.

D) Amazingly enough, not only does the captain not have to land the aircraft to log flight time, and not only is it allowed to be PIC time, but it is also allowed even when he or she is not present on the flight deck. Most incredible of all, a captain *doesn't even have to be awake, or on duty*. He can be snoring away in a bunk (say, while another relief crew is flying), and it still counts as PIC time.

Regulations In Question: Answers

Answer to Question #1: Technically it's B, realistically it's usually A, and in reality the answer that really "counts" is choice C. The only law you're not allowed to break in an emergency is gravity. The exemptions to 91.135 mentioned in choice D have to do only with clearances, communication, and transponder operation, not flying VFR. (Note: It's always a good time to brush up on your "FARs", or check me on my use of them. For those of you searching on the internet, when you get there, use a 'find' in your browser to take you to the appropriate one, e.g. 91.135.)

Answer to Question #2: It's A. The criteria in choice B is for tethered balloons. Answers C and D are bogus.

Answer to Question #3: The answer is D. And what's more, nautical twilight is 12 degrees, and astronomical twilight is 18!

Answer to Question #4: It is choice D. FAR 47.15 (and Advisory Circular 20-27E) cover the subject of special N number requests to the Aircraft Registration Branch (AFS-750) in Oklahoma City. That $10 will reserve a number (if it hasn't been taken) for one year. There are specific format rules. For example, if you're a ham radio operator and your call sign was N6JAZ (like mine), it wouldn't be allowed (but N666JR might be). The FAA responds with Form 8050-64.

Answer to Question #5: It's two of them: A and B. If the maximum cruising speed exceeds 180 knots, it's 12 inches. The same is true if you're going anywhere outside the US or into an Air Defense Identification Zone. (That may change in the not too distant future.) If it's not exactly a real zoomer though, they can be as small as three inches high. (Also the word "Experimental" must be displayed near each entrance in letters that are two to six inches high.)

Answer to Question #6: The answer is B. For choice A, it's actually 200 feet. For choice C, it's really 30 minutes, and the call should be to Flight Service. For choice D, it's 36 inches and 200 feet apart. All the requirements trivia you could ever want to see is in Advisory Circular 70/7460-1K and FAR Part 77.

Answer to Question #7: The most correct answer is, in fact C. And that agrees with a Letter of Interpretation on this one, from an FAA Chief Counsel, in case you still have doubts...

Answer to Question #8: The right one is choice B.

Answer to Question #9: True! Change Notice 3, July 12 2001, contains paragraph 7-6-4, titled "Unidentified Flying Object (UFO) Reports" which gives telephone, fax, and internet access for the single "point of contact" recognized by the FAA for reporting UFO information: the National Institute for Discovery Sciences. They keep a national database on anomalous phenomena, and also periodically share these data with the FAA.

Answer to Question #10: D, folks. They used to be C, but due to an internal conflict between the Federal Acquisition Regulations and the FAA's use of the FAR acronym, what we used to call the FARs is now formally referred to as Title 14, Code of Federal Regulations (14 CFR). This was actually done a number of years ago, in April of 1997. (The CFR catalogs rules adopted by the Federal Government, dividing rules into 50 titles covering all areas subject to Federal regulation.) Title 14 CFR contains rules related to aviation and space. (But it's gonna be quite a spell before all of us stop sayin' "FARs".)

Answer to Question #11: True. Yes, it is theoretically possible, but it would be unlikely. Most amateur radio call signs begin with a letter, many (but by no means all) of which start with the letter "N". They can also start with K or W, or two letters. Most classes of amateur radio licenses end in three letters. (One exception is the amateur extra class, which usually has a two letter suffix. The problem is that Title 14 CFR Part 47.15, sub-part b, states that registration numbers can be made up of from one to five symbols, only the last two of which may be alphabetical. (If either of your initials is an I or an O you're out of luck; to avoid confusion with the numbers 1 and 0, they're not allowed.) There's a special FAA form for reserving N numbers, and it's about $10 per year, until you have an airplane to put it on.

Answer to Question #12: It is choice B. The ability to resist door, bolt, and hinge impact forces of 300 Joules or 221.3 foot-pounds of force, two times, or a 250-pound pull on the handle or door knob for three seconds. (The impact forces are actually simulated by using an approximately 99-lb (45 Kg) weight swung out on a 5.61 foot arc at a horizontal swing distance of 4.48 feet.) These standards are derived from the National Institute of Law Enforcement and Criminal Justice Standard 0306.00 of May 1976, formulated by the Law Enforcement Standards Laboratory of the National Bureau of Standards. They were designed to prevent only incidental or unintentional entrance into the cockpit, and not to thwart the efforts of a "determined person." (See Advisory Circular 25.795-1, dated 1/10/2002.) These standards are being upgraded.

Answer to Question #13: No, it hasn't got anything to do with Groom Lake and flying saucers. That stuff about Area 51 is actually true, but there weren't 240 of them. The other two supposed associations with the number 240 are pure hogwash. But frequent flyers might be interested: it's choice C. Rule 240 applies only within the United States (and an airline often has its own particular version, based on their "conditions of carriage" with the Department of Transportation). It covers you in the event of a delay (which generally must be more than two hours) that is the airline's "fault", such as mechanical problems, a delay in arrival of the airplane or its crew--or a cancellation. It only covers ticketed passengers who arrive on time, needless to say. Most airlines (except, perhaps, America West) are fairly publicity-conscious and are generally responsive. Among the possible benefits of Rule 240 are hotel accommodations (though usually only if the delay is more than four hours and occurs between 10 PM and 6 AM), meals (though sometimes that's only for first class), a brief (three-minute) phone call, etc., when things get out of hand. More commonly, they will arrange a flight for you, on another airline. Note that because the computer systems of different airlines don't all communicate with one another, don't count on your e-ticket being endorsed from one airline to another. A paper ticket might mean a little more money and/or time at the ticket counter, but if you're traveling during busy holiday periods, bad weather, or if there are labor problems or security delays, it just might be worth it. Obviously, it helps here to have only carry-on bags and not be encumbered by previously checked baggage. And the first thing you should do if your flight is cancelled is *not* immediately storm the ticket counter, for re-booking. Instead, proceed quickly to the nearest telephone and either book another flight on the same airline, or on another carrier flying to your destination. *Then* go stand in line. (Unlike the others, you'll already have your flight arranged, as well as your reservation, for endorsement to the new carrier.) Before airline deregulation in the late 1970s, it used to be a Federal requirement. By the way, don't get too snippy or supercilious with this new-found knowledge. First, not all airlines *have* a Rule 240. For those that do, most won't volunteer this information, so it is up to *you* to call them on it. *Do* stand up for your rights, but...do it tactfully!

Answer to Question #14: The FAA (neé Federal Aviation Agency) was "created" in 1958, but the first US aviation laws were enacted by Congress in 1926 (the Air Commerce Act). The International Convention for Air Navigation in Paris in 1919 established basic principles of air sovereignty, required national registration of aircraft, placed restrictions on movement of military aircraft, established a body of basic rules of airworthiness of aircraft and pilot competency, and provided for police measures. Even earlier, in 1907 (October 18, to be exact), the Second Peace Conference at the Hague established an edict (later conspicuously ignored) which prohibited the dropping of bombs from airplanes. Even more amusingly, on July 17, 1908, the town of Kissimmee, Florida passed a statutory regulation which required aircraft registration as well as aircraft "behavior" (speed and altitude) within municipal limits. On June 8 of 1911, Connecticut created regulations governing aviation. Many other early laws, such as the prohibition of flying on Sundays (England) were enacted, as well. (One such law supposedly applied to

Joseph and Jacques Montgolfier soon after their first balloon flights, but it has not been substantiated.) It's C.

Answer to Question #15: According to FAA Order 8700.1, titled "General Aviation Operations Inspectors Handbook", Volume 2, Chapter 1, Section 10 ("Miscellaneous Certificate Information", paragraph 19), the FAA began issuing "blue seal" certificates to those who have "demonstrated their ability to control an airplane by reference to instruments" back in 1961. According to this same reference, a CFI certificate is still just plain black, but can be gold as described in choice C. (Volume 2, Chapter 13 of this FAA Order discusses this.) However, even though the original intent was to serve as an incentive to other pilots to get instrument instruction, choice D is correct. Many private pilots now have blue seals on their pilot certificates. It's D.

Answer to Question #16: As we've probably all been taught in ground school, next in line after an emergency condition comes an "urgent" condition. Although it's rarely done, any pilot can theoretically say the words "pan-pan", repeated three times, whenever he or she is doubtful about fuel endurance, weather, or anything that could adversely affect the safety of flight (and this includes being "momentarily lost" or otherwise apprehensive about knowing one's position). However, that's *not* what is specifically listed next, in the ATC Handbook.

Air traffic controllers are obligated to provide service to aircraft according to duty priorities. The highest priority of ATC service, as we know, is any situation involving an aircraft when an emergency has been declared. (Pilots must abide by this rule, as well as other priorities decided by aircraft category and situations described in sections 91.113 and 91.115 of the Federal Aviation Regulations, otherwise known as the "Right-of-Way Rules".) There is actually a hierarchy of other decreasingly dominant operational priorities in the controller's handbook. (Although there can be many factors involved, no standard list can be guaranteed to uniformly apply in every situation, and an individual controller should exercise his or her own judgement from a safety standpoint, there *is* such a list.) And there's *way* more than four more. According to Chapter 2, paragraph 2-1-4, these operational priorities are:

a) aircraft in distress
b) civilian air ambulance ("LIFEGUARD" flights) or military air evacuation flights
c) search and rescue (SAR) flights
d) presidential aircraft
e) Flight Check aircraft (flight inspection of navigational aids, direction DF equipment, and RADAR)
f) NIGHT WATCH (or NAOC, Natl. Assoc. of Ordnance & Explosive Waste Contractors) aircraft
g) FLYNET aircraft (nuclear, biological, or chemical accident emergency response teams)

h) Garden Plot (CARF) aircraft (military support related to domestic civil disturbances)
i) aerial sampling (SAMP) aircraft (atmospheric sampling for nuclear contamination)
j) interceptor aircraft
k) special air mission and law enforcement operations (SCOOT)
l) NOAA mission aircraft
m) IFR aircraft
n) OPEN SKIES (treaty observation or verification) SUA flights
o) aircraft operating under the National Route Program
p) diverted flights

So the "correct" answer is choice C. Incidentally there are also two additional situations besides emergencies, under which ATC is excused from the normal procedures and minimums contained in the controller's "bible", FAA 7110.65: except when deviations are necessary to conform with ICAO documents or special agreements involving ATC service outside the US, *or* other procedures dictated by a military document, FAA directive, or a letter of agreement. That's what it says.

Answer to Question #17: It's A, and although it's not as critical, B as well. The two forms mentioned in choice A look very much alike; the former states that it has that dual function, and it also has a "DD" at the top. Don't laugh! This kind of mix-up has happened! If just the "medical" is issued to a student pilot applicant, it means that he or she could be flying without proper certification, and could be held responsible, if the FAA finds out. If the wrong box is checked in Block 62 (you get the right FAA form, but the wrong box is checked), it means that no record will exist that you are a student pilot. You wouldn't receive possibly critical safety information, nor would you have a "pilot's access" to DUAT online. Obviously, the AME should be the first one to check against this sort of error. (By the way, there is no such form as the FAA Form 8055-5 mentioned in choice D.)

Answer to Question #18: Surprisingly, it's actually spelled out fairly well. Part of the apparent confusion may be that its apparent optimism and leniency may seem too good to be true. It isn't. In this case, it's in CFR Title 14, Part 61.23(c)(3)(ii)(A). According to the rule, a third class medical certificate (one issued after September 16, 1996 at least, which does seem kind of academic, by now) expires at the end of the 36th month after the month of the date of the examination shown on the certificate, if the person has not reached his or her 40th birthday on or before the date of the examination. In fact, according to the AOPA ePilot, Volume 3, Issue 24, on June 13, 2003, even for someone who is 39 and will shortly be 40, their medical certificate is valid for 36 calendar months--the last day of their "birthday month", three years later. The answer is choice B.

Answer to Question #19: There are quite a few instructor ratings, such as: Airplane Single Engine, Airplane Multi-Engine, Rotorcraft Helicopter, Rotorcraft

Gyroplane, Glider, Instrument Airplane, Instrument Helicopter, Powered-Lift, and Instrument Powered-Lift. (There aren't many of *those*, yet.) And despite what you'd think, a CFII doesn't have to be a CFI, first; these ratings are all totally independent of one another. There is *no such thing* as a basic instructor rating (the well-known "CFI") that is a mandatory first step to being any *other* kind of a flight instructor. One doesn't necessarily have to start as a CFI in a single-engine airplane, for example. Or to be more esoteric, CFI applicant can initially qualify as an instrument powered-lift instructor. As retired FAA Aviation Safety Inspector, Designated Pilot Examiner, and attorney Frank Phillips, Jr. wrote in a recent issue of "FAA Aviation News", each CFI rating stands alone. So the answer to the question is: "false" and although it might indeed *be* better instruction, it might *not* be, and it also might not be *legal*! The bottom line is that, although there are a hierarchy of pilot certificates, as well as ground instructor certificates, and the vast majority of flight instructors working in single-engine trainers probably *did* work their way up the food chain, as one would expect, it isn't necessarily always going to be so...

So, check!

Answer to Question #20: The biennial flight review must consist of at least an hour of flight time. The Pilot Proficiency Award Program is usually three hours, for fixed wing pilots...unless you happen to be going for your Wings in a glider. In that case, it can theoretically be under an hour, although they would have to be rather short flights to meet the quota as prescribed in Advisory Circular 61-91H, Section 7, paragraph (d). What most pilots never pay attention to however is the following: In that same Section 7, it also requires a *fourth* hour if the pilot is applying for an odd-numbered award phase, *and* he or she is not instrument-rated (and also, even if instrument rated, if that pilot is not also instrument current). Thus, even though the "Wings" card has only three lines for the separate hours of flight training required, as described in AC 61-91, only current instrument-rated airplane pilots can actually get away with those three. If you're not in that group (unless you happen to fly helicopters, gliders, airships, or ultralights), you need four. The answer is choice D.

Answer to Question #21: Theoretically any aircraft that moves at a relative speed through the air which is great enough to allow a "readable" ribbon of useful size to remain extended behind it could do the job. But the only ones commonly used to convey aerial written messages (other than the moving marquees sometimes seen on the sides of a blimp) are airplanes and, believe it or not, helicopters. Section III of the FAA publication FS-I-8700-1, "Information For Banner Tow Operations" does mention the obvious considerations involving a helicopter's tail rotor, and 14 CFR Part 133 covers some these issues as well. As for airships, they don't need to actually tow anything, with all that advertising real estate on their rumps. The answer is C.

Answer to Question #22: Generally, as it's specified in CFR Title 14, Part 91.155, VFR visibility minimums in Class G ("uncontrolled" airspace) increase from one statute mile during the day to three miles at night. There is one exception: You need only one statute mile of visibility at night, and need to simply remain "clear of clouds", if you stay within a half a mile of the runway (and you're below 1200 AGL). As soon as you go beyond that, weather minima become three statute miles and 500 feet below the clouds (plus 1000 above, and 2000 feet horizontally). But within this short reach, nighttime VFR weather minimums increase to their daytime counterparts. (Be aware though, that the moment you inadvertently enter a cloud, you'd be in violation of the rules. Clouds are much harder to see at night, so this is a real concern. Just because it's permissible, it isn't necessarily safe.) You don't have to make your landings to a full stop unless you need to maintain currency for carrying passengers at night. Since you've met most of these criteria, but not all of them (being more than a half mile from the runway on the downwind leg, for one thing), the answer is choice A.

Answer to Question #23: It's D. Here's why. First, yes, the U.S. Terminal Procedures publication does state: "...If it is necessary to maneuver at speeds in excess of the upper limit of a speed range for a category, the minimums for the next higher category should be used." (A descent is a maneuver too, incidentally.) True, you'll find many approaches where the minimums are indeed lumped together for categories A and B, but by no means does that preclude the need to know the answer to this question. If you were coming out of the clag on the NDB approach at Coeur D'Alene, Idaho, that 100 foot difference in MDAs might be important! But back to the somewhat academic question: nowhere in Title 14 CFR does it give a definition of the imperative "should". (It never says "must" or the even more definitive "shall".) That said though, remember this: When you fly faster, you'll need more room for obstruction clearance because of an increased turning radius, and lower altitudes for a lower category might not give you as wide a swath of maneuvering room. Also, fly faster, and you need more reaction time, which eats up more distance (so your visibility minimums should be greater, as well). Incidentally if you ever forget which category means what approach speeds, just look at a few approach plates. You'll usually see maximum aircraft speeds for circling approaches shown in lieu of approach category letters. (This started with charts dated after July 27, 1989, and with good reason, as circling approaches are more hazardous and the less uncertainty as to what's safest, the better.) You probably won't see approach category E on any approach plates. That's for aircraft with approach speeds over 165 knots!

Answer to Question #24: D. This came out in the Fall of 2003. You still need the presence of an "authorized instructor" as mentioned in CFR Title 14, Part 61.51(g)(4). Incidentally, that means that he or she must be either an Instrument Ground Instructor or an instrument flight instructor (a CFII). The term "authorized instructor" is defined in Part 61.1, sub-paragraph "b". I was able to confirm the newly conferred status of this type of PCATD with a phone call to the somewhat blustery Aviation Safety Specialist and Operations Inspector within the FAA's AFS-840, General Aviation and Commercial Division, Certification Branch, in Washington, DC. And of course, if your flight school doesn't happen to have a brand new Advanced PCATD, the old ATC-610J, with all its glorious and anachronistic op amps and integrators, has long been legal for helping you maintain instrument currency (though you still need an "authorized instructor" looking over your shoulder, figuratively if not literally).

Answer to Question #25: It's B. You would have to jump through a few procedural hoops, but actually, a Part 91 pilot and his or her airplane can attain Category II approval without needing any special avionics and instrumentation, manuals, or a copilot. What sort of hoops? The first one is writing a formal "letter of intent" to your local Flight Standards District Office requesting a waiver from the rules requiring a copilot described in CFR Title 14, Part 91.189(A)(i), as well as those requiring a manual (91.191), and the *l-o-n-g* list of avionics alluded to in 91.205, sub-paragraph (f) (although the list itself is in Appendix A to Part 91). In it you would list your airplane's avionics, that they are all certified, and explain that you already have an instrument rating, that both you and your airplane are current for IFR flight, and that you do not fly for compensation or hire (the one limitation to all this). At this point, you aren't actually *mailing* the letter, just writing it. Then you get a copy of FAA Form 7711-2, the Application for Certificate of Waiver or Authorization (which is what you'll need to earn the actual Certificate of Authorization For Certain Category II Operations, Form 7711-1), fill it out, and attach it to your letter of intent. You then request a meeting with a FSDO general aviation inspector, to whom you show the letter, as well as Volume 2, Chapter 59 of FAA Handbook 8700 (which states what he is duty-bound to do, in order to help you get your waiver). Next, you take a check ride with an FAA inspector, demonstrating your ability to hand-fly an ILS down to 100 feet. (One warning: if your airplane is a twin, you'll also have to do it on one engine.) If you manage all that, you'll be granted approval to bust Category I as far down as 150 feet. Do three more practice approaches with a safety pilot down to 100 feet, write another letter affirming these to whomever signed your initial authorization, asking for 100-foot DH (and also perhaps, 1200 foot RVR minimums), and that's it. Simple, huh?

Answer to Question #26: False. They actually do so on a fairly regular basis; they just don't *sound* like IFR holds. Whenever Approach Control tells you to hold "North of the freeway, and call back in five minutes" those <u>are</u> holding instructions. If you've ever flown to Oshkosh in late July, you already know about VFR holds.

Answer to Question #27: It's choice (C). ATC will block the airspace for 30 minutes before allowing in another IFR or SVFR aircraft inside it (FAA 7110.65 10-3-1 and 10-4-1). In the meantime however, phone calls will be made to the

FBO, the airport manager, local law enforcement, or anyone else who can go and look for the aircraft.

Answer to Question #28: Choice (A) is true. So is choice (B). In fact, although it's rather unlikely that a general aviation airplane has another internal space that would be very far *from* the flight deck, according to Title 14 CFR, Part 61.51(e)(3), an authorized CFI can log as PIC time all of the time during which he or she is acting as the authorized instructor (which is usually from an adjacent seat). Choice (C) is fiction. That leaves...choice (D). According to FAR 61.51(e)(1), although a recreational, private, or commercial pilot may log pilot-in-command time only for that flight time during which that person—

> (i) is the sole manipulator of the controls of an aircraft for which the pilot is rated.
> (ii) is the sole occupant of the aircraft, or
> (iii) except for a recreational pilot, is acting as pilot in command of an aircraft on which more than one pilot is required under the type certification of the aircraft or the regulations under which the flight is conducted...

That's not so for an airline transport pilot (which is what an airliner's captain would probably be). FAR 61.51(e)(2) says that an airline transport pilot may log as pilot-in-command time all of the flight time while acting as PIC of an operation requiring an airline transport pilot certificate.

Chapter 12: Things That Make You Think

Experience isn't simply a library of what has *happened* to you. A truer and more complete definition of experience would argue that it is actually also what you have *thought about* what has happened to you. It is only when you reflect upon your training (or any encounter with something new) that can you truly integrate that experience into a greater whole. Although I'll admit that the path leading to the doorstep of sublime understanding may lead you through a few vacant lots along the way, it does have its rewards; it is this process which elevates rote learning to the realm of wisdom. In this chapter for example, you will learn more about precession and magnetic variation and how they affect flight instruments, and about the workings of magnetos. You will discover how birds deal with wind, and perhaps most immediately (in the very first among this bunch) you'll be able to hammer a stake through the heart of that perennial nuisance regarding bathtubs and yes, even the proverbial kitchen sink.

Subject: The Chronosynclastic Infundibulum

Question 1: Does a bathtub or sink REALLY drain clockwise in the Northern hemisphere, and counter-clockwise down under?

A) Absolutely. I saw it on the web. (Sinks work better, though.)
B) Coriolis force can be apparent on smaller scales only when there are absolutely no currents or internal eddies within the body of fluid. If you let the tub settle for about an hour or so, then could somehow make the plug vanish without causing any disturbance, yes it would drain clockwise in the Northern hemisphere and the other way in Southern latitudes.
C) No. It's a myth. You'd need a bathtub the size of a stadium.
D) Actually, in a totally stable fluid body, there *is* no spiral motion towards any opening at the bottom.

Subject: More Deer In the Headlights:

Question 2: Your throttle linkage just broke. You're stuck in climb (or cruise) power. What are your options?

A) Bail out.
B) A very hot and fast landing.
C) Use carb heat, put the wheels down, switch to one mag, lean the mixture, turn on all electrics, trim up a bit for slower speed, climb, and mayday!
D) Everything in C, but also head for a nice two mile runway, then cut the power over the numbers, and just let it slow down and settle in sweetly.

Subject: Inertial Navigation Systems

Question 3: Which is the most astoundingly correct statement with regard to inertial navigation systems?

A) A freely mounted gyro-stabilized platform solves the problem of misinterpreting pitch changes (such as on rotation, at take-off) as rapid accelerations, but if it is correctly aligned with the earth's surface at departure, it would effectively report an outside loop as the aircraft travels (possibly halfway around the world) to its destination. Therefore, although they are not "nailed down", the gyros do require continual alignment with (or references to) the local vertical.
B) Their sensitivity is so great that a freely mounted gyro platform (if left powered up) would appear to tilt over the space of a few hours, even if the aircraft remained motionless and parked on the ramp, due to the earth's rotation alone.
C) Initialization of an INS takes longer at higher latitudes because it takes the gyros longer to sense the earth's rotation and find true North.
D) all three

Subject: Up, Up and Away

Question 4: What is the highest altitude to which a non-rocket powered aircraft could ascend?

A) about 85,000 feet
B) about 100,000 feet
C) about 123,500 feet
D) about 170,000 feet
E) over 350,000 feet

Subject: "Which Way Do We Go, II"?

Question 5: When one is at a high latitude, it is well known that a magnetic compass is, at best (unless one is intimately familiar with local idiosyncrasies), nearly useless. What is another type of simple early navigation device which can in fact be used to find direction at any time during the day, even at the poles?

A) Aside from your Mark I eyeballs, there is none.
B) Two magnetic compasses, several miles apart, can be used to infer location based on known lines of magnetic flux.
C) the sun compass
D) a cross staff, or mariner's astrolabe

Subject: Attitude Readjustment Hour

Question 6: In most general aviation airplanes, the standard "six pack" configuration of instruments includes an attitude indicator and a heading indicator, both of which are vacuum powered. Most of us know that every so often, we need to manually readjust our heading indicator against the magnetic compass to correct for precession. Yet when you think about it, aside from an occasional pre-takeoff adjustment, we hardly ever need to adjust the attitude indicator in flight. Why is that so, and in what circumstances might we want to think about doing so?

A) Attitude indicator gyros would indeed precess just like our heading indicator, if they spun at the same speed, but because they spin over ten times faster, their "rigidity in space" is much more persistent. Unless there is turbulence that is severe or extreme, they are required by CFR 14 Part 23.1303 not to precess more than one degree per hour. This is more than adequate for all but ocean-crossing flights.
B) Attitude and heading gyros both precess at the same rate, but attitude indicators actually have two gyros spinning in opposite directions, and each precesses such that the opposite movements cancel out. This does not immunize them against the effects of turbulence, which introduces nutation (a "nodding" of spin vectors in addition to precession) and can result in accumulated errors.
C) The vertically oriented axis of rotation in attitude gyros allows a means of automatic correction with the addition of a pendulum which senses the local gravity vector. This either opens air blast vanes on an air-driven gyro or powers electric torquer motors to return the spin axis back to vertical when it precesses.
D) The best answer, as usual, is the most simple and elegant one: the spin axis of an attitude indicator is locked in the vertical and thus does not precess.

Subject: Dippy Compasses

Question 7: The earth's magnetic field runs parallel to the earth's surface at the equator, but in northerly latitudes it dips downward "into" the earth. This is what is responsible for the familiar turning errors, introduced in ground school. To a very rough first approximation, what are the magnetic dip angles at 30 degrees and 60 degrees North? (The field is not radially symmetrical about the poles, and so there is also a longitudinal influence.)

A) one and a half degrees down at 30N, four degrees down at 60N
B) three degrees down at 30N, eight degrees down at 60N
C) 12 degrees down at 30N, 18 degrees down at 60N
D) 50 degrees down at 30N, 75 degrees down at 60N

Subject: Hot and Short

Question 8: Out of all the public use runways in the US, if you also factored in the density altitude, using a normal standard atmospheric lapse rate, where would the "effectively shortest" runway be located?

A) Telluride Regional Airport, in Telluride Colorado (TEX), has one runway (09-27) that is 6870 feet long. Its 9,078 foot altitude however (even on a cool "standard temperature" day of about 26.5 deg F) would seem to be almost three times shorter (for a normally-aspirated airplane) at about 2300 feet!
B) The Lake County Airport at infamous Leadville Colorado (LXV), at 9927 feet, is the highest public use airport in the US. Its one 6400-foot runway 16-34--it also has a 100 x 150 helipad--is, due to its altitude (which even on a "standard day" might be a chilly 23.6 degrees F) would seem to your Skyhawk to be only 1900 feet long! That's well under one-third of its actual length, and that would be on a "good" day!
C) Colorado, yes, but not Leadville: Runway 03-21 at McElroy Airfield (elevation over 7400 feet MSL) n Kremmling, CO, is just 1100 feet long. Although its other runway (09-27) is 5540 feet long, even with a standard temperature for that altitude (which is just above freezing, at about 32.5 degrees F) runway 03-21 would, to a normally aspirated airplane, feel like it was only 469 feet long.
D) Erstwhile Field, 1500 feet from the summit of Pike's Peak, is the world's most treacherous airport. Only turbocharged STOL aircraft or turbine helicopters are allowed to use its 1900-foot runway, which in even the cool thin air (about 20 F deg below freezing, on a standard day at that altitude) would seem over six times shorter, at 290 feet! The runway perimeter fence is made of aluminum. (Guess where they get it?)

Subject: Murphy's Corollary: Headwinds

Question 9: True or False: On almost any given round-trip cross-country of one or two hundred miles done on the same day, for all practical purposes, any wind is a headwind.

Subject: Maybe Not Even Steven

Question 10: For many Cessna singles, why will fuel burn NOT be the same between the two tanks when the fuel selector is on "Both"?

A) Manufacturing differences, which may include slight variations in fuel line lengths and drain forces, can favor one tank over the other by several percent.
B) Flying just slightly wing low can effectively cancel the equal draw of the "Both" position. (In this case, the higher tank will empty faster.)
C) There is no bias, trim-induced or otherwise. "Both" means both!

D) The fuel vent, located below the left tank, is the culprit.

Subject: Sliding Home

Question 11: You're about to enter the traffic pattern in a retractable gear aircraft, and when you lower the gear, much to your consternation, you don't see three greens. After trying everything possible, including switching bulbs, checking circuit breakers, sudden pull-ups, performing the manual extension procedure, and a tower fly-by, and after you had tried everything else you could think of (or sought advice for over the radio, from folks on the ground) the gear is still up, and it does not want to come down. Which of the following surfaces would be best to land on?

A) gravel or sand
B) water
C) a foamed runway
D) concrete or asphalt, hold the foam
E) grass

Subject: Landing "Downwind"

Question 12: For what reason might you want to land on the downwind side of a runway, in crosswind conditions?

A) There is no possible advantage. Like the useless runway behind you, doing so would leave little wiggle room if you forfeited the runway surface on your leeward side, for if, despite your best efforts, the wind blew you off to the side of the runway, you would then already have one wheel almost in the weeds. Not only that, but landing on the windward side of the built-in transverse slope (even though it's only a degree or three) also confers some small "wing low" advantage into the wind.
B) If the runway were wide enough, landing on the leeward side and then following a slight diagonal path towards the windward side, a few hundred feet away, would to a small extent lessen your crosswind component during the critical first seconds before the airplane slowed enough for most of its weight to be transferred from the wings back to the wheels.
C) Whenever there is a heavy rain combined with a crosswind, the built-in transverse slope (higher in the middle than on the sides) might actually work against you. The crosswind can cause the water to build up on the windward side, against the runway's convex slope, and the increased depth on that side could conceivably increase the likelihood of your landing gear to begin hydroplaning.
D) During a snowy crosswind, the build-up of possible snow will occur first on the upwind side of the runway. Also, the runway surface is usually less smooth away from the center, and this improves traction.

Subject: Slip Sliding Away...

Question 13: If the tire pressure on your airplane is 25 psi, at what speed could hydroplaning occur?

A) 45 knots
B) 50 knots
C) 54 knots
D) 63 knots

Subject: Hometown Sluggers, 9...Search And Rescue, 1 ??

Question 14: Many distress alerts that are sent from the NOAA's Sarsat Mission Control Center in Suitland Maryland over to the Air Force's Rescue Coordination Center at Langley AFB, Virginia are false alarms. Although there is some reduction in their number by the time the AFRCC initiates activity with the Civil Air Patrol (or also, the Coast Guard), about 75% of the distress signals received by the AFRCC are not actual beacons, but are instead transmissions from devices such as:

A) ATM machines
B) pizza ovens
C) stadium scoreboards
D) all three, and many more

Subject: Highway Tolls

Question 15: In general, it is more expensive to fly distances of up to a very few hundred miles (via commercial airliner, on an individual basis, that is) than it would be to make the same trip by car. For trips of more than this, it is much more efficient to fly there, in terms of fuel and energy expended per passenger-mile. From the following choices of the percentage pairs of the world's transportation energy budget that is consumed by air travel, and that which is needed for highway travel, which comes closest to reality? (Note: the two will not add up to 100%, as there are other forms of transportation, such as over water, which are not considered here.)

A) air, 25%; highway, 40%
B) air, 20%; highway, 50%
C) air, 15%; highway, 60%
D) air, 10%; highway, 70%

Subject: Beyond the Third Degree

Question 16: We've all heard of the three axes commonly discussed regarding an aircraft's attitude and resulting movement in three dimensions: pitch, roll, and yaw. But actually when it comes to flight training devices, there are actually three more components of motion. What are they?

A) lateral, longitudinal, and vertical acceleration
B) heave, surge, and sway
C) quite simply: up, down, and sideways
D) spin, lift, and frequency

Subject: Why Two?

Question 17: You may have noticed that certain aircraft have more than one pitot tube. Why?

A) CFR Title 14 for Parts 121, 125, and 135 operations (Subpart C, Aircraft and Equipment) require it, for the sake of redundancy during icing conditions (and both must be heated).
B) Large aircraft produce greater disturbances in airflow requiring more than one collection point in the air stream
C) birds
D) It is required for all aircraft that require more than one pilot.

Subject: Discombobulated

Question 18: Which of the following situations is most difficult for a pump-driven attitude gyro to overcome?

A) a sustained turn at a shallow bank angle
B) two or more 360-degree steep turns
C) due to the loads placed on the gymbals, just one single snap roll
D) in a gyro without a caging mechanism, a single loop, because it can damage the bearings
E) a long descent at idle speed

Subject: Weird Wings

Question 19: Which of the following innovations never existed:

A) wings that folded in flight
B) wings that changed their angle of incidence in flight
C) wings that changed their camber or shape in flight
D) wings that telescoped in flight
E) wings that inflated
...or...
F) none of the above

Subject: The Magneto March: Left, Left, Left-Right-Left....Left, Left, Left-Right-Left...

Question 20: Perhaps you've noticed that key-operated magneto switches are configured to read, from left to right: "OFF-RIGHT-LEFT-BOTH-START" (or something like that). Why is "left" on the right, and "right" on the left? Why isn't it "OFF-LEFT-RIGHT-", etc?

A) It's not even an issue; not all are that way. It actually depends on who made the switch; many indeed are also labeled as "OFF-LEFT-RIGHT-BOTH-START"
B) Because that's the way it was always done; i.e., it was purely arbitrary.
C) Because each relative position (not its letter label) *probably* reflects which respective p-lead is grounded.
D) The magnetos are intentionally wired as they are viewed from the front of the airplane, and not from inside the cockpit.

Subject: Magic In Numbers, II

Question 21: A pilot can use the face of the directional gyro to determine the proper succession of headings (if there is no wind) or perpendicular ground tracks in order to fly rectangular patterns. If the pilot is familiar with the area, local landmarks often provide assistance. So do judging consistent distances and angles from the runway. Though it has been somewhat over-publicized, there's also supposedly another trick. What is it?

A) Always use the exact same crab angle for each leg of the pattern.
B) The distance from the runway to the downwind leg should always be one half mile for each 60 knots of airspeed.
C) Turn base at a 45 degree angle for zero wind, and 10 degrees sooner for each part of a headwind that equals one-fifth of your airspeed on final.
D) Adding the digits of each no-wind heading (whether it is the upwind, crosswind, downwind, base, or final) will always add up to the same number.

Subject: Not *Enough* Of A Good Thing

Question 22: Non-pressurized flight above 14,000 feet MSL without supplemental oxygen creates obvious problems for pilots. Aside from reduced aircraft performance and other commonly known problems experienced by non-turbocharged aircraft at higher altitudes, a non-pressurized normally-aspirated Cessna 172 that *could* manage to claw its way up that high would suffer from what other problem?

A) a rough-running engine if the pilot doesn't lean sufficiently
B) a *non*-running engine, if the pilot leans excessively
C) a rough-running engine due to timing problems with the magneto
D) loss of control from only slight increases in load factor or turbulence

Subject: For the Birds

Question 23: When birds "come in for a landing" what is the major factor influencing their landing direction?

A) obstacles
B) Birds almost always land in the same direction in which they have been flying, regardless of wind.
C) Just like us, it's wind; birds almost always land into the wind to minimize their forward speed and maximize the braking action afforded when they suddenly spread their wings to catch the wind and stop.
D) There are no determining factors.

Subject: More Stuff For the Birds: No DF Steers For Feathered Friends

Question 24: Birds, like pilots, don't always have a tailwind. As a matter of fact, they have the same problems we do. How do they handle crosswinds in flight?

A) They continuously crab while en route.
B) They drift off course, and make a series of corrections as they go.
C) trick question: Birds do neither. Actually, they get blown off course, and then when in sight of their destination, they "home in" on it visually. In cases where greater distances are involved, they use the Earth's magnetic field to do the same.
D) They get lost, then walk the rest of the way.
E) We have no idea.

Subject: What's Wrong With This Picture?

Question 25: Well, what ***is*** wrong with this picture?

photograph of the vertical stabilizer from the Airbus A-300 from American Airlines Flight 587, which crashed in Jamaica Bay in New York on November 12, 2001.

A) nothing
B) The rudder is missing.
C) The flag is backwards.
D) all three

Subject: Chinese Water Torture

Question 26: An airplane with leaky fuel caps is sitting out in the rain. How long might it take each drop of water, the moment it gains access to the fuel tank, to: A) reach the bottom of the tank; and b) get to the engine?

A) a few minutes to reach the bottom of the tank, and a few hours more to reach the engine
B) a very few seconds to reach the bottom of the tank, and a few minutes more to reach the engine
C) a few minutes to reach the bottom of the tank, and only a few seconds to reach the engine
D) just a few seconds to reach the bottom of the tank, but a few hours to reach the engine

Subject: Down And Around

Question 27: From the North Pole, one can only go southward. Not being the trusting type, rather than following the slightly suspicious-looking sign pointing every which way with directions and distances to several of the world's major cities, you decide to set out on your own. Unfortunately, you forgot your watch.

The problem is that, to know where one is going, one must know the meridian being followed. Without a way to tell time, how could one determine one's longitude?

A) easily, with a nautical or astronomical almanac
B) by using a Jovilabe
C) by examining the shadow of any object stuck vertically in the ground
D) wait until dark

Subject: After You, My Dear Alphonse!

Question 28: You're taxiing your airplane at an airport in Great Britain. The taxiway, like all other taxiways nearly everywhere, has a yellow stripe down the middle, and that's just where you are. Your dilemma arises when you realize that there is another airplane headed toward you moving in the opposite direction, several hundred yards ahead. The taxiway is wide enough to accommodate both airplanes, but of course, that yellow line *isn't*. Now you begin to wonder...what are you supposed to do? You remember that in the UK, one drives on the left, but you also know that regardless of that, airplanes are expected to taxi down the center line of a taxiway, in whatever country that may be. (The same is true for runways, obviously.) Back home, you reason, where you drive an automobile on the right, but where you've always passed slower vehicles on the *left*, you begin to wonder whether or not the other pilot will expect you to pass to his right, or to *his* left. You realize this may be silly...or is it? After all, in the UK, where they drive on the left, they certainly don't *taxi* on the left! Isn't passing to the right of opposing traffic a standard procedure? So, deary me, what would you do?!

A) Even in ICAO (International Civil Aviation Organization) countries where automobiles drive on the left, such as Great Britain, such things are governed by ICAO rules of the "road." British pilots operate on taxiways exactly the same way we do. You would each pass to the right.
B) Believe it or not, there *are* no fixed rules covering this situation. Turn towards whichever side offers the most room.
C) In ICAO member countries where driving is on the left, passing is also on the left: left rudder!
D) Each of you would stop, queue up along with the *other* airplanes waiting in line to pass, and work this out in a civilized fashion.

Subject: Don't Sweat the Small Stuff?

Question 29: True or false: Pilots deal better with minor in-flight problems than they do with more critical malfunctions.

Things That Make You Think: Answers

Answer to Question #1: An infundibulum is a funnel-shaped cavity, from the Latin, derived from *infundere*, "to pour". (Then there's that *other* even stranger word from that strange expression, which comes from the "The Sirens Of Titan" by Kurt Vonnegut, and has to do with different kinds of truths connecting together agreeably. I just threw that in for show.) But the answer is C. On this smaller bathroom-sized scale, frictional and viscous forces greatly eclipse Coriolis forces, and a draining bathtub is really just cyclostrophic flow (a balance of pressure gradient and "centrifugal" forces). When we see clockwise flow around an atmospheric high (up in the Northern hemisphere) we are really looking at a broader category called "geostrophic" motion--again a balance of pressure gradient and Coriolis forces. Coriolis forces (an apparent force, really, as it cannot *cause* a motion) only show up in nature beginning at a scale of several hundred meters. This might be compressed down to dozens of meters in a fluid with absolutely no "vorticity", in a controlled chamber. My intuition tells me also that it would "work" better using a fluid that was less dense than water. However, my guess is that this wouldn't make very good grounds for a funding request, though.

Answer to Question #2: The answer of course is D.

Answer to Question #3: It is D.

Answer to Question #4: It's D. 85,000 feet was the *approved* ceiling of the SR-71. 100,000 feet was the probable maximum altitude of Helios, the experimental solar-powered airplane (which reached 96,500 feet before it was destroyed in an accident in June 2003). The world altitude record for a jet airplane is held by a MiG-25 Foxbat at 123,524 feet. And the right answer, D, is 170,000 feet: the theoretical maximum altitude of a *very* large hydrogen balloon. The X-15 attained 354,200 feet back in 1963, but it was rocket powered. (Of course, that was recently eclipsed by Burt Rutan's SpaceShipOne, also rocket powered, on October 4, 2004, when it flew its X-prize winning second flight to an altitude of 367,442 feet, and it and others will doubtless continue to push aircraft ceilings ever higher.) Yes, this one was a meanie: five choices!

Answer to Question #5: It is choice C. The Sun Compass draws on the fact that the sun's shadow from a point (like a sundial's gnomon) in the middle of a disk describes different hyperbolas at different times of the year. You don't have to know the time of the day in order to get general directions within a few degrees. It is believed that the Vikings may have used a similar "bearing dial" instrument. (At high latitudes, as elsewhere, the direction of the sun is towards east in the morning and towards the west in the evening, but during the middle of the day of course, its location in the Southern sky is much less ambiguous than at lower latitudes.)

Answer to Question #6: C. There really is such a design, and it works. However, even though the system is not disoriented by coordinated turns, very shallow-bank turns, or flying out of trim for long periods (due to sloppy flying or improperly rigged controls) can introduce errors.

Answer to Question #7: The right answer is choice D, believe it or not. The approximate formula is $tan(\boldsymbol{\delta}) = 2 \cdot \tan(\boldsymbol{\lambda})$, where $\boldsymbol{\delta}$ = dip angle, and $\boldsymbol{\lambda}$ = latitude.

Answer to Question #8: Choice D, although extrapolative fiction, would be true in principle. (But there isn't really an airport up there.) Out of all of the approximately 5000 public use airports that were in a recent AOPA database, Leadville Colorado was the highest. But it is Kremmling, choice C, that takes the cake. There were also about 14,000 private airports (give or take a few), for which we do not have length and elevation data, and some of these may very well be higher (or certainly shorter), and thus even worse.

Answer to Question #9: It's true. This would probably not apply with a frontal passage or a significant shift in wind velocity between the initial and the return trip, but for a simple scenario, any tailwind during one half is going to be more than cancelled out by the headwind during the other, because of the greater time that headwinds act on the airplane overriding what I call the "shorter tailwind time". If you do the math for some "standard" empirical winds as a fraction of true air speed (TAS), the general formula would be, solving for time from the relationship of Distance = Rate * Time (where D = one-way distance, and 2D = round-trip distance):

$$2D / TAS \; < \; ((D / (TAS - wind)) + (D / (TAS + wind)))$$

Regardless of whether you're flying a Lear or a Cub, whether the winds aloft are halacious or just a gentle breeze, and whatever the relative angles are between you and the wind, the following examples will hold true, in regard to how it will lengthen your trip (compared to no wind at all, which is an extreme rarity):

wind = 1/100 * TAS = only 1/9999 longer
wind = 1/20 * TAS = only 1/399 longer
wind = 1/10 * TAS = only 1/99 longer
wind = 1/8 * TAS = only 1/63 longer
wind = 1/5 * TAS = only 1/24 longer
wind = 1/4 * TAS = still only 1/15th longer
wind = 1/2 * TAS = hmmm... 1/3 longer
wind = 70.7% of TAS = yipes! TWICE as long!

wind = 3/4 of TAS = whoa! 2 & 2/7th times as long!

The disadvantages in terms of increased trip times start out small, but in the extreme case of the wind equal to your TAS of course, the disadvantage is infinite.

Answer to Question #10: The answer is D, but an explanation is in order. Older Skylanes in particular (prior to about 1979) had this problem. Even with the fuel selector on "both", pilots have often seen one tank (in this case, the left) seemingly empty faster than the other. Whenever fuel is used up, it needs to be replaced with air, lest a vacuum form inside the tank, which could lead either to interruption in fuel flow to the engine, or in the case of bladder tanks, simply deformation as they get "sucked dry", or worse, with non-bladder tanks, structural deformation. First, there is an "L" shaped vent tube behind the left wing strut, and this is connected to the upper outboard section of the left wing tank. There is a vent interconnect line between the upper inboard areas of both left and right tanks, to equalize pressure between the two. But when the tanks are full---and the slight dihedral of even a high-winged airplane doesn't help here---this interconnect line can actually be "submerged". As fuel is used, air coming in from the vent tube pushes fuel through to the right tank, replacing fuel as it is used from it. (This can even continue as fuel drains below the level of the inboard interconnect tube via the occasional sloshing over the interconnect port.) This has been seen to occur even to the point that the left tank is reading only 1/3 full, and the right tank is still reading "full". No fuel starvation will occur however, as the engine is still indeed accessing both tanks--although fuel exhaustion is never less likely if you let them both run dry! (Another nagging problem these vents cause is of course overflow, if you fill the tanks with cold fuel and the fuel warms enough, as the temperature rises.)

Answer to Question #11: The best answer is choice D. It would also be advisable to take your sweet time and burn off as much extra fuel as you could, within reason. Using a hard surface runway causes the least damage to the aircraft, and usually no injuries to those on board. Foamed surfaces are actually too slippery and also give poor deceleration and can easily result in a loss of directional control. Grass, sand and other soft runways (as well as landing on a body of water) are liable to cause parts of the aircraft to “dig in”, resulting in violent deceleration or sudden directional changes, all of which greatly increase the chances of injuries and structural damage.

Answer to Question #12: This one may not be so obvious. However, I *did* say that some of these would make you think, now didn't I? It's C. Choice A of course makes sense in most cases (such as when it isn't raining heavily). One might also argue for choice B. But choice C is worth remembering when the weather is really lousy. (You might wind up asking yourself again though, why it was that you'd wanted to land there.) Choice D may have some theoretical truth, but it's mostly irrelevant.

Answer to Question #13: It would occur at the speed given in choice A: Just like with a car, as you may have heard (but filed away in a dusty cabinet of your memory, hopefully ready to recall when needed) the speed at which hydroplaning can occur is, as weird as it sounds, nine times the square root of the tire inflation pressure. The square root of 25 is five, and five times nine equals 45.

There are several reasons that it's a good idea to keep this particular trivia item in mind. First of all, tire pressures can vary greatly between different airplanes that you may fly, and even with the same airplane, it can still vary. For a Cessna 172 for example, if you walk out and the tires look soft and squishy--that's a technical term--with a bulge on the bottom, and you suspect (hah!) that they're under-inflated, you just might have only 25 psi in there, or even less. On the other hand, if you walk out and the tires look rock hard, round all around, and not all of the tread is in contact with the pavement, it might be as high as 50 psi. Of course you're supposed to keep the tire pressures at the level that the manufacturer recommends, because even pressures that are only 10 psi low would allow your airplane to hydroplane at a lower speed--which you may not want, one rainy day.

Several things besides the tires can also influence hydroplaning. Tire treads having grooves that allow water to escape from being trapped between the runway and the rubber would obviously increase the allowable depth of water before hydroplaning could occur. (Depending on tread design, such a critical water depth can range from one to four-tenths of an inch.) The more worn the tire, obviously the more likely it will hydroplane. For you heavy iron drivers out there, undercarriage assembly also affects hydroplaning characteristics. With tandem "bogies" there is a kind of clearing action from the *front* wheels, which reduces the depth of water encountered by the *rear* ones, therefore delaying hydroplaning and increasing the given depth of water before you and your passengers start taking water skiing lessons. Runway surface is obviously also just as important. Surfaces that have a coarse matrix (or better yet, grooves) will take longer to build up to limiting depths, especially in strong winds. With little or no wind, most runways have an adequate transverse slope for good drainage during all but the heaviest of rain. However, drainage can still be compromised with a crosswind above 10 knots, because a crosswind that is blowing up that slope could allow water to build up higher on that side. This would lead to hydroplaning on that side, or (if your airplane was big enough to straddle the hump) asymmetric braking. Very heavy precipitation rarely lasts long, and waiting a few minutes is usually enough for water to drain off sufficiently to get rid of really critical water depths.

Answer to Question #14: Would you go for choice D? I hope so, because that's the right answer; if you guessed all three, you would be correct. Although many false ELT signals are caused by events such as hard landings, malfunctioning ELT units, etc., actually according to Dr. Ajay Mehta, the Sarsat Program Manager with NOAA in Suitland Maryland, with whom I have spoken, around three-quarters of the alerts sent from their Mission Control Center are not actual beacons. This is because EPIRBs and ELTs operate in a very crowded neighborhood, frequency-wise.

Answer to Question #15: If you guessed D, well then, that was a lucky guess. It's actually more like nine percent and about 72%, respectively. So even though over 40% of the initial weight of that 747, in which you and a few hundred other folks are zipping along at maybe 500 knots, is just from fuel, and perhaps only five percent of the weight of your car is due to the gas you had to pump into it, flying is still cheaper (at least on a demographic usage level) overall.

Answer to Question #16: Ahoy there! It's B. The familiar roll, pitch, and yaw are angular. Heave, pitch, and sway however are linear. They are simply terms for displacement and lateral motion. Together they make up the six "degrees of freedom" in the motion we experience. These terms are used both in reference to ships at sea (or as in the case of this example below, ducks in the bathtub), as well as the motion platforms on which some of us get to ride, aka: flight simulators. Surge is "forward and back" along the longitudinal, or roll axis. (Just think of "surging forward".) Sway is sideways, along the lateral, or pitch axis. (Just think of swaying palm trees, if that helps.) And heave is "up and down" along the vertical, or yaw axis. (Perhaps the image of a "heaving deck" on rough seas might help.) More bubble bath, anyone?

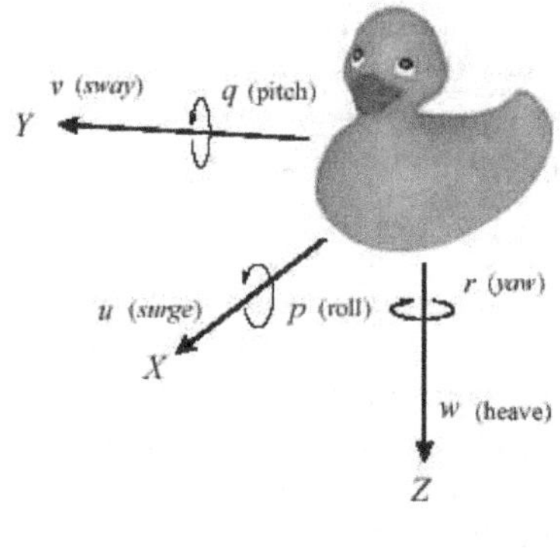

Answer to Question #17: Yup, birds. It's C. It's a rule in the "FARs" all right, but it's in Paragraph 1323 of Parts 23 and 25 for airplanes. It says that where duplicate airspeed indicators are required (such as with commuter and transport category airplanes) their respective pitot tubes must be far enough apart so that one bird can't take them both out. (Parts 27 and 29 for rotorcraft don't seem to care.) Section 23.1323 of Advisory Circular 23-17A also *advises* that for Part 135 operations, the minimum acceptable straight-line distance between them is 14 inches. (This might account for a four-pound bird.)

Answer to Question #18: It's choice A. As has been mentioned in another trivia question, currently the attitude and heading indicators in most general aviation airplanes are most often vacuum powered. Although we need to periodically readjust our heading indicators against our magnetic compass to correct for precession, we rarely need to make in-flight adjustments to our "artificial horizon" because our usual vertically oriented axis of rotation allows a means of automatic correction with the aid of a pendulum which senses the local gravity vector, allowing this instrument to erect the spin axis back to the vertical. This is somewhat akin to the phenomenon of low accelerations going unnoticed within the semi-circular canals of our own inner ears. Although from a mechanical standpoint, abrupt changes in g-loads *are* hard on the gimbals, with sustained shallow turns, it is also almost impossible for an attitude gyro not to be fooled into giving an erroneous reading. Remember that the next time you perform any series of successive practice course reversals under the hood! Typical approaches usually are not flown long enough, or at a low enough power setting, to be a problem. (However, high altitude flight or excessively low temperatures can create trouble.)

Answer to Question #19: We've all seen some pretty wacky ideas when it comes to wings, whether that means their shape, configuration, or (especially during the earliest years of aviation's first century) their number. Other questions have discussed multiplanes, variable-geometry wings such as those of the Messerschmitt P-1101 jet prototype or the more modern "swing-wings", forward-swept wings, and even ornithopters. Even the Wright Brothers' "wing-warping" idea was an innovation, although now it may seem quaint. But if it was possible, it's likely that someone, somewhere, has tried it. (Well, it's likely, but it's not absolutely certain--hence the question).

There have been wings with anhedral (negative dihedral), inverse taper (wider at their tips than the wing roots), and some designs have even involved so-called "Magnus flow" over wings that would simply consist of rotating cylinders (but that wasn't one of the choices). In a way, many common aircraft involve using wings with variable camber (choice C); Fowler flaps and leading edge slats already do that. Research aircraft such as the F-111/MAW (mission adaptive wing) have gone a step further, using integrated leading and trailing edge segments of flexible fiberglass which were internally articulated and allowing completely variable camber along the full span of the wings. (Actually, NACA's H.F. Parker proposed a true variable-camber wing in a 1920 paper.)

There have also been variable angle of incidence wings (choice B). Back in 1931, one George Cornelius tried a variable-incidence wing aircraft and followed it a few years later with his "Mallard," which had both variable incidence *and* forward-swept wings. Variable incidence wings of course exist in powered lift aircraft (like some tilt-rotors), and also were involved in early designs such as the Boeing Vertol Model 76 (aka VZ-2A) in 1957, or the Canadair CL-84 in the mid-sixties. The Martin XB-51 also had variable-incidence wings (and in both of these last two, the tail plane tilted to match, as well). The F8U Crusader tried a variable incidence wing. The entire leading edge and ailerons could also be lowered to increase wing camber and reduce approach and landing speeds, *and* also rotated about its rear spar to raise the whole wing (which improved the pilot's visibility). The outer wing tips could also fold for carrier storage.

And speaking of folding wings, the short-lived XB-70 Mach 3 bomber had wingtips that would lower *in flight* by 25 degrees above 300 knots, and down to 65 degrees when above Mach 1.4. One of the most extreme examples of a folding wing was in fact a biplane whose entire lower wing folded in flight, making it into a monoplane! (So much for choice A.) By about 1940, almost every country (except Italy and the Soviet Union) had given up on the production of biplane fighters, as designers couldn't compensate for the slower speed caused by the drag of a second wing. However, the Soviets had a novel idea: in about 1940, the Nikitin-Sevchenko design bureau (V. Sevchenko & V.V. Nitkin) built a series of aircraft called the IS-1, IS-2, and IS-4 (for Istrebitel Skladnoj, or folding fighter) in which the whole lower wing was designed to retract into the upper wing, in flight, and *after* the landing gear had retracted into the lower wing! It had a cantilevered gull wing anchored in the top of the fuselage, forward of the cockpit. The lower wing had an inverted gull shape. For takeoffs and landings, the bottom wing was extended and locked, but once airborne the pilot could crank up the hinged lower wing, aided by hydraulics. The inner wing panels, into which the wheels were retracted, fit upright inside wells in the side of the fuselage. The outer panels bent at a right angle to the inner ones and also fit into wells, but in the underside of the upper wings. Thus, it was more than a clever compromise between a biplane and a monoplane fighter: it was both. In a way, it was an early form of variable geometry wing. In its biplane configuration, it had excellent takeoff and landing ability, as well as increased payload, while as a monoplane, it had both a higher speed as well as greater range.

As for inflatable wings (choice E), that's nothing. Would you believe an *entire inflatable airplane*? In 1956, Goodyear designed and built several (a dozen, actually) prototype inflatable aircraft. It was called the XAO-3 "Inflataplane". The single-seat version had a wingspan of 22 feet, cruised at about 60 mph, and it had a range of about 390 miles. The two-seat version was a bit bigger and slower. (Of course, powered parachutes and paragliders can be said to have "inflatable" wings, too...)

Then there's the telescoping wing idea: This was another contender for an early type of "variable-geometry" aircraft; it was built in France and made its first flight in August of 1931. It was designed by Mikhail Makhonine, a Russian engineer. The MAK-10's outer wing panels actually retracted into the inboard panels. A low wing monoplane, its wings were each made up of three sections. The first one was made of metal and was built into the slim fuselage. The second and third were made from wood, and could be moved on rollers to telescope inside the metal section. With wings fully extended (from 43 to 69 feet and with a surface area increase from 226 to 335 square feet), although its top speed did drop by about 30 mph (from about 155), its range increased from 6220 to 7480 miles. The concept of variable-span or retractable wing airplanes has also since been further explored by the military, individual entrepreneurs in other countries including the US (e.g., the Gevers "Genesis" aircraft), academia (Virginia Tech for example), and other organizations (such as the AIAA), whether for improved tactical versatility or simply "roadable" aircraft. The advantages of such versatility are, at least in theory, alluring. With its shorter span, along with reduced drag and higher speed would come greater wing strength, an improved ride in turbulence, lighter stick forces at high speed, natural protection from icing on the outboard wing sections until needed for landing, it could simulate higher performance jet aircraft for advanced pilot training...and it could fit in smaller hangars! With its wings stretched, in addition to greater endurance and range, it would also have an improved climb rate, better handling and control at low speeds, a higher ceiling, better glide ratio, lower takeoff and landing speeds as well as ground run...such a deal! So what's the catch? As usual, probably a combination of developmental inertia, sparse infrastructure, the economy, demographics, and engineering complexity that isn't obvious to anyone but engineers. But scratch duck "D" in this line-up.

So the answer is: choice F. (Yes, it was intended as a double negative; *none* of the above *never* existed.)

The Makhonine, with wings extended (top), and retracted (bottom), from a 1931 French journal, courtesy of the Archives Division of the National Air and Space Museum reprinted with permission from Archives Division of NASM

Answer to Question #20: A magneto is both a low voltage alternating current generator *and* a high voltage coil and distributor, providing for an aircraft what the battery/alternator, high voltage coil, and distributor do, on a car. Magnetos act as a generator and create their own low or primary voltage, then act as a spark coil and transform the low voltage into high voltage. The high voltage spark gets routed to the spark plugs at about the right time to ignite the fuel-air mixture in each cylinder. So magnetos convert rotational energy into high-voltage pulses that are used to fire your aircraft's spark plugs, without the need for external power from a battery or the electrical system. Because they work even when there's a total electrical failure, they are a preferred spark source, for obvious reasons. The "left" position is positioned on the right and the "right" position is labeled on the left because this is one way to indicate which of the respective P-leads is grounded (more on that in a minute). What's a P-lead? The P-lead is a thick shielded wire (about 16-gauge) running from the cockpit ignition switch to the ungrounded end of each magneto coil's primary winding. (The "P" stands for "primary" by the way.) Its sole purpose, in a way, is to actually "break" something. Well, it allows the ignition switch to *disable* a magneto (or both of 'em) by grounding the "hot" side of the primary coil in the rotor of the magneto. (The rotor is like a cam operated spark generator. A spark is generated because of a sudden--and intentional--collapse of a magnetic field in one coil and the induced and stepped up voltage--the spark--in another coil.) When the P-lead is grounded through the ignition switch, these so-called breaker points are *unable* to interrupt the current flow in the primary coil, making the magneto incapable of generating a spark. Maybe it helps to think of those dangly strips you see on some automotive vehicles, which allow stray static electricity "sparks" to be *grounded*, and get harmlessly dissipated, down to the ground. By the way, I may be covering old material here, but for those new to all this stuff, aircraft engines have *two* spark plugs in each cylinder and *two* magnetos. One plug in each cylinder is fired by one magneto. The other magneto fires the *other* set of plugs. The benefits of this "dual ignition" are that if one system fails the engine will still operate, and also that there is some improved combustion of the mixture (with two plugs firing in each cylinder), which increases power output and performance. Okay then... (I must warn anyone who is dyslexic: you're gonna have a *real* tough time on *this* one.)

Taking as an example the Bendix 10-357200 series ignition switches, if the ignition key is in the "R" position, the *left* lead is grounded (so that it *can't* spark), and the right mag is *un-grounded*, so that it *can*. (The common way that folks refer to this "enabled" state is that we say the mag is "hot".) When the ignition switch is in the "L" position, the *right* lead (mag) is grounded and the left mag is the one that's ungrounded, or "hot". (To help you along a bit here, just think of a kid who has misbehaved, and who has then been grounded by his parents: he can't leave the house...or circulate...or make any more sparks for awhile...does that help?) So thank goodness for small favors; the letter on the ignition switch at least represents which mag you're "on". Why they reversed their position in terms of "handedness" would make an interesting study in terms of the lineage of functionality, but my guess is that it was their way of indicating that you're grounding the mag other than the one you selected. So...to reiterate: when you rotate the ignition switch to that left position (which happens to be *labeled* "R"), you're grounding or shutting off the left mag; when you check the mag that is labeled "L" (which, as I have already beaten to death, is on the right), you're grounding the "right" (not wrong!) magneto. And, finally, when the switch is on BOTH, you are un-grounding both mags. So...the position *label* on the ignition switch is for the mag that's activated, while the positions themselves are for the ones that *aren't*. So...the answer's C...for "Clear as mud"...right? (Notice by the way that I used the word "*probably*"...you may also come across different ignition harnesses, or other wiring diagrams like this one, which in this case seems to show that whichever mag you select means that's the one you're grounding out! The bottom line is, as you always hear, don't move a prop unless you're well positioned to get outta the way in a hurry, 'cause the mags could be "hot".)

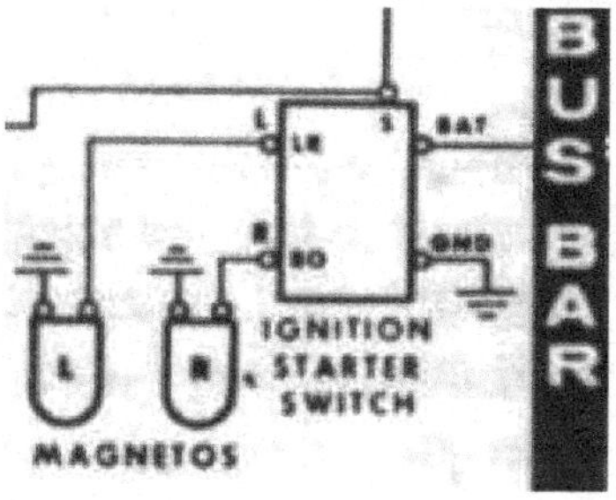

Answer to Question #21: D. Choice A would make no sense at all (unless the wind direction was exactly 45 degrees from the runway). Choice B? I made it up (as if you couldn't guess). Choice C looks like it might make sense, actually, but I made *that* one up, too. On the other hand, choice D, now that one, I *didn't* make up. Actually, this so-called "amazing discovery" was supposedly first made (maybe "realized" is a better word)...by a twelve year-old boy! The story goes that while riding in a J-3 Cub flown by his father in New Mexico, Justin Winfield noticed that the digits for the headings flown on each leg had a curious relationship: when added together, he noticed that they added up to the same number! He parlayed this observation into winning third place at his school's science fair, and then won an FAA award at the state level, as well as an Air Force Gold Medal Award. (I'll remain circumspect and withhold any thoughts on the obviously discriminating criteria for profundity on the part of certain science fair judges.) But as an example, let's say you take off from runway 13, and in a normal pattern, turn a left crosswind. Your no-wind heading is 040 degrees. Turning a left downwind, it becomes 310. Your left base would be 220, and final of course, 130. In each case, lo and behold, the digits add up to four. That so-called amazing fact might've astounded the teachers in New Mexico, but it's not exactly an undiscovered property of a base ten number system wherein adding nine to any two-digit number *de*creases the ones digit by one and *in*creases the tens digit by the same amount (up to 36, after which, at least for runways, it recycles). And in a similar way, *subtracting* nine from any two-digit number *in*creases the ones digit and then *de*creases the tens digit, again both by one.

Of course, it's easy for right-handed patterns where the "90" gets added to your heading (or multiples of nine get added to the runway number), but wouldn't you know it, most of our patterns go the *other* way. The same principle does work in reverse of course, but it's just a bit trickier.

There's only one small problem with this brilliant rule. It doesn't always work. Try it with a runway 01, 02, 10, 11, 19, 20, 28, or 29, and things get a little confusing (if that's all you're paying attention to, that is). Of course maybe he knew that and there's a "modulo nine" clause in Winfield's Rule. Here is a table of two sets of headings for upwind, crosswind, downwind, base, and final legs, with both right and left-handed patterns for runways in a clockwise sequence starting with 36 and back to 36 again. Notice what happens to this, using certain runways. I think I'll stick to landmarks and pilotage to keep my *own* path over the ground on the straight and narrow...

Right-handed Patterns:					SUM:	Left-handed Patterns:					SUM:
360	090	180	270	360	**9**	360	270	180	090	360	**9**
010	100	190	280	010	**1 or 10**	010	280	190	100	010	**1 or 10**
020	110	200	290	020	**2 or 11**	020	290	200	110	020	**2 or 11**
030	120	210	300	030	**3**	030	300	210	120	030	**3**
040	130	220	310	040	**4**	040	310	220	130	040	**4**
050	140	230	320	050	**5**	050	320	230	140	050	**5**
060	150	240	330	060	**6**	060	330	240	150	060	**6**
070	160	250	340	070	**7**	070	340	250	160	070	**7**
080	170	260	350	080	**8**	080	350	260	170	080	**8**
090	180	270	360	090	**9**	090	360	270	180	090	**9**
100	190	280	010	100	**1 or 10**	100	010	280	190	100	**1 or 10**
110	200	290	020	110	**2 or 11**	110	020	290	200	110	**2 or 11**
120	210	300	030	120	**3**	120	030	300	210	120	**3**
130	220	310	040	130	**4**	130	040	310	220	130	**4**
140	230	320	050	140	**5**	140	050	320	230	140	**5**
150	240	330	060	150	**6**	150	060	330	240	150	**6**
160	250	340	070	160	**7**	160	070	340	250	160	**7**
170	260	350	080	170	**8**	170	080	350	260	170	**8**
180	270	360	090	180	**9**	180	090	360	270	180	**9**
190	280	010	100	190	**1 or 10**	190	100	010	280	190	**1 or 10**
200	290	020	110	200	**2 or 11**	200	110	020	290	200	**2 or 11**
210	300	030	120	210	**3**	210	120	030	300	210	**3**
220	310	040	130	220	**4**	220	130	040	310	220	**4**
230	320	050	140	230	**5**	230	140	050	320	230	**5**
240	330	060	150	240	**6**	240	150	060	330	240	**6**
250	340	070	160	250	**7**	250	160	070	340	250	**7**
260	350	080	170	260	**8**	260	170	080	350	260	**8**
270	360	090	180	270	**9**	270	180	090	360	270	**9**
280	010	100	190	280	**1 or 10**	280	190	100	010	280	**1 or 10**
290	020	110	200	290	**2 or 11**	290	200	110	020	290	**2 or 11**
300	030	120	210	300	**3**	300	210	120	030	300	**3**
310	040	130	220	310	**4**	310	220	130	040	310	**4**
320	050	140	230	320	**5**	320	230	140	050	320	**5**
330	060	150	240	330	**6**	330	240	150	060	330	**6**
340	070	160	250	340	**7**	340	250	160	070	340	**7**
350	080	170	260	350	**8**	350	260	170	080	350	**8**
360	090	180	270	360	**9**	360	270	180	090	360	**9**

table of runway numbers and results of applying "Winfield's Rule", by the author

Answer to Question #22: The cold, thin air can also contribute to arcing and cross-firing in unpressurized magnetos, due to the fact that the insulation value of air is reduced along with its density. Air is a pretty good electrical insulator. Take it away, and the sparks won't fly where and when we want them to. (Pressurized magnetos have sealed caps and plugs and obtain the needed extra pressure from a turbodischarger device.) So aside from mixture issues and mushy controls, the answer is choice C: you would also get caught in the crossfire from your own engine.

Answer to Question #23: The right answer is C. Only a birdwatcher can prove it: watch some, sometime. There are exceptions though, such as when a large bird like an albatross has to land on a small island in strong crosswinds, in which case it will crab, and then make a "crosswind landing".

Answer to Question #24: First, the long-winded (pardon the insufferable pun) version: This is one question which actually doesn't have a simple answer; the subject still generates controversy and discussion among students of bird migration. Typically, a migrating bird is observed at one place and time, and knowing the wind at its altitude, scientists can determine the bird's course and ground speed, as well as its heading and air speed. What they usually don't know is the bird's destination (lacking flight plans, and all), which leaves them guessing as to whether it's experiencing wind drift from a fixed heading, or whether it's adjusting its heading (and/or its speed) to compensate for the wind (however completely) in order to "fly direct". With satellite tracking however, it is possible to follow large birds over their entire course of migration. Because many birds exhibit a high degree of fidelity to specific nesting or migratory destinations, once a given individual has been followed for a complete "there and back" cycle and enough data has been accumulated, there will be better answers to the question of where it was trying to go.

The answer is, at least for the present, that birds are sometimes drifted completely by crosswinds, and sometimes *correct* completely. Larger, faster-flying birds, such as waterfowl or shorebirds, are less subject to drift than slower birds (like small songbirds). In order to assess the amount of drift, most birds need some visible ground reference. Birds flying over large bodies of water appear to experience wind drift. They may use waves (also however influenced by the wind), but this seems to result in only partial correction for the drift. There is evidence that birds may allow themselves to be drifted or to compensate for drift depending upon where they are in their migratory journey. The closer to their destination they are, the more likely they are to correct because precise navigation at that point is obviously important. Earlier in the trip, drift experienced during one flight might be compensated by drift in the opposite direction on another flight, or lighter winds on a subsequent flight might make compensation less costly. This stuff can all be subjected to optimality modeling, and to some degree, the birds do seem to behave according to the predictions of the models.
And the short answer: Choice E. We have no idea (yet).

Answer to Question #25: This is a photograph of the vertical stabilizer from an Airbus A-300 that was American Airlines Flight 587. It crashed in Jamaica Bay on November 12, 2001. Among the suspected possible causes were severe and sudden side loads imposed by repeated rudder movement, wake turbulence from the JAL 747 that preceded it, bogus parts, structural failure, and engine failure. However, aside from the missing rudder (as well as the rest of the airplane no longer being attached), there is nothing wrong with this picture, and it hasn't been fooled with. While I do not intend to trivialize the tragedy and loss of life that resulted from this accident, there is something else that you might notice, if you looked closely. Depending upon how much one knows about the proper ways in which the American flag is supposed to be displayed, this could be nothing new, or else a bit startling. The flag really *is* "backwards". (Even reversing the photo would of course not change its relative orientation.) On the left side of any vehicle or aircraft displaying the American flag, it is supposed to be depicted with the "canton" or "union" (the blue field with the stars) in its familiar upper-left position. However, on the starboard side, by convention it is always displayed with the stars "facing front". This is done so that the flag looks as it would if it were blowing in the relative wind created by forward movement. This also applies to flag patches worn on the sleeves of many military uniforms, incidentally. For example, there are two separate patches in the Army inventory: the normal flag replica worn on the left, and what is referred to as the "reversed field" flag patch, which is worn on the right sleeve. However some organizations (for example, the Boy Scouts) do not follow this convention. I don't know how this might apply to a Harrier hover-taxiing backwards, but the answer is A.

Answer to Question #26: As you probably know from draining a fuel sample from a wing tank of an airplane (either from first-hand experience, or vicariously), a slug of water will immediately make its way to the bottom of the fuel strainer. Of course, this is because water is heavier than avgas, and given a free path, the difference in densities means a rapid trip to the bottom for any parcel of entrained water. So that knocks out choices A and C. However, that slug of water, in addition to being allowed to escape into your fuel strainer or GATS jar when you visit the fuel drains, can also get to the fuel line, and make its way to your engine! There's some good news, and some bad news: because water (and all liquids) are basically incompressible, and because of the adhesive forces in a narrow fuel line, that water is sitting atop a column of fuel and cannot easily make its way into your engine. After perhaps six, eight, or maybe ten hours though, it *can*. So if you don't drain your fuel system well at the "engine" end, and you just drain the quick drains and take off, you may have just enough time for a run-up and takeoff. The owner's manual for a Cherokee 235 says to first drain the fuel tanks, and then the belly drains from the tip and main tanks (11 seconds for each tip tank, in fact, because that's how long it takes to drain them). When you drain the tanks like that, you also pull out any water that got into the fuel line, and might already have made its way to your engine! So the answer is D. This information was provided to me by the good graces of CFII Pat Olivolo of Maryland. As a former FAA

Aviation Safety Inspector, he has personally investigated at least two accidents that occurred under circumstances such as those just described.

Answer to Question #27: Latitude is easy, of course. Longitude isn't. Columbus couldn't tell how far East or West he was without a means of reckoning time. A century later, Galileo invented the Jovilabe, which was an instrument for determining an observer's longitude (without the use of time) by observing the rotation of Jupiter's satellites. Celestial navigation works too, of course; Saint Exupery used the Southern Cross as a guide to navigating down the coast of Africa, and kept it on his left while flying up the coast of Africa to Europe (though I have a feeling that the coastline was kind of handy). Acceptable answers are either or both choice B and D. Of course, if it's much past the middle of March, there *is* this small problem that you could be waiting about six months before it got dark enough to find your way via the night sky...

Answer to Question #28: It's A.

Answer to Question #29: Actually, they don't. You can read a paper on that very thing in the October 1995 issue of the *Flight Safety Digest*, published by the Flight Safety Foundation (not AOPA's <u>Air</u> Safety Foundation, but actually a separate independent, nonprofit, international organization, headquartered in Alexandria, Virginia). In this study, Captain Robert L. Sumwalt and Captain Alan W. Watson of the Battelle Aviation Safety Reporting System analyzed different aircraft malfunctions and how effectively their flight crews handled them. In their report, "ASRS Incident Data Reveal Details of Flight-crew Performance During Aircraft Malfunctions," their findings suggested that responses to less serious malfunctions are associated with more "error-chain" symptoms and adverse consequences. This is because in dealing with serious malfunctions, pilots are usually trained so that their responses become reactive and preconditioned (thus avoiding the "error chain" syndrome). Minor problems often involve more reflective responses (plus, they are not rehearsed as much in training) and can result in a distracting element, even to the point where situational awareness is compromised or possibly lost, and the frequency of errors was several (about five) times greater. The answer is: "false".

Chapter 13: Other Things You'd *Never* Think About

The best and most memorable way to present any subject matter is to make it entertaining. Throughout this book, I have to admit that I didn't need to try too hard to inject a bit of humor and whimsy into the mix, because I've always found aviation so remarkably endowed, all at once, with both our most extraordinary aspirations and our most noble failings. Still, I've tried to maintain some passable level of decorum. This last chapter though, is probably going to raise the most eyebrows. Here's where I get to stand up on the podium, put on my checkered jacket and bow tie, and present a few of aviation's sideshow curiosities. Who were the *other* Wright Brothers? How might you find your name on an aeronautical chart? What were Otto Lilienthal's last words? Did you ever wonder what would happen if your car had the same "six pack" of flight instruments as many small airplanes? What are the odds of experiencing a simultaneous electrical and vacuum failure, while flying in instrument conditions? Did the US Air Force ever try building a flying saucer? How many soda cans you could make from a Cessna 172?

I thought you'd never ask.

Subject: Now For One On GPS:

Question 1: What change in GPS would triple the accuracy available to the "user segment", but has been dropped by the Air Force?

A) End-of-week rollover errors caused too many problems with the computers in Cheyenne Mountain, so the US government shortened the more precise of the two pseudo-random code sequences from over 38 weeks to precisely 12. However, this cut the precision by a factor of three.
B) WAAS
C) something called crosslink: The GPS IIF "follow on" series was originally supposed to have the ability to relay nav updates to the "control segment" via relay vehicles, which would have greatly reduced the "age of data" (from a day to three hours or so) of each constellation vehicle's own position, and thus would improve signal accuracy to the user. Alas, it's not to be. The Air Force dropped it from their specs due to funding cutbacks.
D) Huh?

Subject: The "Other" Wright Brothers:

Question 2: Most people (even most pilots) don't know *this* one. In addition to Orville and Wilbur Wright, there were actually two <u>other</u> Wright Brothers near the turn of the 20th century who were, in a way, connected with aviation's infancy. Who were these men?

A) William and Robert Wright, the landlords of the building in which Orville and Wilbur had their bicycle shop. (They were stepbrothers, incidentally.) They were not related to "The" Wright brothers.
B) Jedediah and Mordecai Wright, also brothers (although they were twins), purchased the printing business that Orville and Wilbur sold prior to their initial investment in the new and then-booming bicycle business in 1893.
C) Milton Wright and his wife Susan actually had two sons before the birth of Orville and Wilbur. Reuchlin was 6, and younger brother Lorin was 4 when the "elder" of the famous duo, Wilbur, was born in April of 1867.
D) In 1899, Wilbur wrote a letter to the Smithsonian Institution inquiring about aeronautical research. Coincidentally, it was answered by a man whose last name also was Wright. He, too had a younger brother, however his name was Orrin, not Orville.

Subject: Famous LAST Words:

Question 3: What were Otto Lilienthal's last words?

A) "My shoe fell off!"
B) "Oh, s******* !" (In German, it's got eight letters.)
C) "Sacrifices must be made."
D) "You are standing on my arm, hollow-head!"

Subject: Excusez-moi, Madame!

Question 4: The term for the angle that a wing chord line makes with its fuselage is called the angle of incidence. Is there a similar term for the angle of the horizontal stabilizer? If so, what is it?

A) Yes there is, and it is simply the negative angle of incidence.
B) Yes, sort of. The term is *decalage*, which is actually the difference in angles of incidence between the wing and horizontal stabilizer.
C) No, there is no such word.
D) Yes, and the term is *decolletage*.

Subject: The People's Airspace?

Question 5: China has over one billion people (actually about 1,340,000,000 as of early 2012). Out of all those folks, roughly how many hold the equivalent of a private pilot license?

A) a little over 200,000
B) about 20,000
C) roughly between 1500 and 2000
D) two

Subject: The Unthinkable

Question 6: What does SCATANA stand for, and to what entity does it direct control of the National Airspace System?

A) It stands for Security, Control, Air Traffic Association National Advisory, and the FAA still runs the show.
B) It's actually an acronym and a contraction, standing for Secure Command Authority, Tactical Air Navigation, and the controlling agency becomes the US Air Force.
C) Security Control of Air Traffic and Navigation Aids, and it's NORAD.
D) It doesn't actually stand for anything.

Subject: Air Conditioning For Eskimos?

Question 7: An airliner is cruising along at FL 350. The captain announces to the passengers that the outside temperature is minus 40 degrees Fahrenheit. In order to keep the cabin nice and comfortable, what system is essential?

A) In almost all airliners, surplus power from the engines is used to power generators that supply electrical energy to generate heat.
B) Actually, due to the friction generated from flying over 0.5 Mach, airliners actually need air conditioners to cool the air before it enters the cabin.
C) Choice A is close, but it is simply excess heat from the engines which passes through radiators with vanes that are used to warm the incoming air.
D) Simply pressurizing the cabin generates heat, and in fact, temperatures could conceivably climb to Death Valley digits if the air were not also refrigerated after being compressed.

Subject: Bogeys, Twelve O'clock? There's Nobody There!

Question 8: What is the squawk code 5200?

A) This is actually used for the F-117A and the B-2 in training maneuvers.
B) This is the emergency code used by the US Air Force for emergencies. (The Army uses 5100, the Navy 5300, the Marines use 5400, and the Coast Guard, 5500.)

C) If you ever hear someone mention a squawk code of 5200, bring your radio into the shop, because you're picking up maritime chatter. This code (and those in the 5200 range) are used by ships at sea.
D) This code is used in the agricultural application industry, much like 1200 is used by the rest of the VFR pilots in the US.

Subject: Unfortunate Terms

Question 9: What is excrescence drag?

A) with respect to transport category aircraft, the drag created by waste fluid seepage from onboard lavatories through septic outlets used in ground handling, and which can freeze at altitude, building up to sizeable chunks of "blue ice" that adheres to the fuselage and creates significant drag
B) a drag factor due to cavities, grooves, steps, ridges, and other surface discontinuities that create non-uniform pressure distribution
C) the frictional drag of an airship expressed as a flat plate cross-sectional area
D) There is no such term.

Subject: Your Name Up In Lights

Question 10: Where might one find one's name on an aeronautical chart?

A) The FAA is running out of identifiers, and there are a number of additional reporting points listed in the National Flight Data Center which have some fairly common names, such as ROBERTS or JOHNSON.
B) at an airport: Many airports have common names, and the odds are actually fairly good that if you have a common name, there is already a public or private airport using it.
C) By looking in FAA publication 7350.7, LOCATION IDENTIFIERS.
D) For a $36 fee, the ATA-100 group of the FAA at 800 Independence Avenue in Washington will send you a CD with every navigational fix, airport identifier, navaid, reporting point, and route, and you can find out.

Subject: Just DUAT

Question 11: What is the origin and significance of the familiar four-letter FAA-sponsored automated weather and flight planning service with which we have all become so familiar?

A) The acronym originated when Data Transformation Corporation, the provider which fielded the first version to the FAA in 1986, suggested that it be named the DTC Universal Access Telnet, or DUAT.

B) An employee at Contel ASC, the forerunner of its subsequent owner, GTE (which was in turn the predecessor of the one following, DynCorp, and which was again renamed when acquired by Computer Sciences Corporation) suggested a name during a software development meeting: DUAT. However, software developers being the way they are, it was actually in honor of Duat, the Egyptian world of the afterlife.
C) It was a corruption of the expression "Do It!" popularized by Nike, and also suggested by a member of an original development group (perhaps in an overinspired moment of cameraderie and team spirit).
D) Direct User Access Terminal

Subject: A Six-pack On Your Dash?

Question 12: If anyone were eccentric enough (as well as possessed of enough skill) to augment their car's dashboard instrumentation with an additional "six-pack" suite of fully functioning analog flight instruments, how well could they be made to work?

A) They wouldn't. Automobiles lack a vacuum system, and of course there would be nothing to supply dynamic and static pressures.
B) The difference between the car's 12-volt system and the airplane's 14-volt system (in addition to the much higher amperage required in an airplane) would render the gyroscopic instruments useless.
C) Cars don't travel extensively enough nor rapidly enough in three dimensions for any such instruments to give meaningful indications.
D) They could indeed be installed, aside from logistical problems with existing dashboard real estate. One would need to install a vacuum pump, and of course a pitot tube, static port, and associated plumbing. But the readings in a car would be just like what you would see in a taxiing airplane.

Subject: Body By Coke?

Question 13: The average 1970s vintage Cessna 172 has enough aluminum to make the equivalent of roughly how many typical soda cans?

A) 250
B) 2,500
C) 25,000
D) 250,000
E) irrelevant. The alloys are totally different.

Subject: Good Help Doesn't Come Cheap

Question 14: The cost of producing one Boeing 777 is approximately the same as that of building

A) an entire shopping mall
B) a mid-size high school
C) a 40-room mansion in Beverly Hills
D) a B-2 bomber

Subject: That's Okay; I Still Have My *Other* Ear...

Question 15: The noisiest aircraft of all time, during takeoff, is (was)

A) a B-1 in full afterburner
B) Concorde
C) the XB-70 Valkyrie supersonic bomber
D) the SR-71

Subject: Oooh, Dad! Can I Get One Of Those For My Bike?

Question 16: The world's smallest jet engine (whether turboprop, turbofan, or turbojet) used in a "real" passenger carrying aircraft weighs in at

A) 19 grams
B) five pounds
C) 75 pounds
D) about 225 pounds

Subject: Improbability Drive, And the Worry Wart

Question 17: For an instrument rated general aviation pilot, a true nightmare would be to experience a simultaneous double failure of both the electrical and vacuum systems, while in IMC. Just how often does something like that occur, and approximately what are the odds of it happening to you?

A) infrequently: It occurs less often than once every 1000 flight hours.
B) remote: Its probability is on the order of greater than one ten-millionth, but less than or equal to one one-hundred-thousandth, per hour of flight, or per event (e.g., takeoff, or landing).
C) improbable: It has a probability of occurrence greater than one one-billionth, but less than or equal to one one-hundred-thousandth, per hour of flight, or per event.
D) extremely remote: The probability of occurrence is between the orders of one one-billionth and one ten-millionth, per hour of flight, or per event.
E) extremely improbable: The probability of occurrence is on the order of one one-billionth or less per hour of flight, or per event.

Subject: Okay, Okay, So Merry Christmas Already. Just Make It Snappy, Pops!

Question 18: Certain fables in the imagination and fragile minds of a young child are best left untouched, until such time as when they reach a certain age that only a parent can know, when gullibility wanes, worldly wisdom dawns, and the time then comes to guide our wee ones along the bumpy path of a sometimes harsh reality. Most men of the cloth seem to have this innate skill, borne perhaps of uncommon compassion, but there was one stand-in vicar in England who achieved unwelcome fame when he explained with somewhat indiscreet harshness just what might happen to Dear Old Saint Nick and his faithful reindeer, if they truly *did* have the necessary velocity to deliver their gifts of joy to every home in Christendom during this one blessed night. Let's just say that he did not break it to them gently. Shocked parents and terror-stricken children listened on as he continued his explanation, thinking the children would really like to know these facts, and he described just how and why Santa, his reindeer, and sleigh would have to be traveling so fast that they would be instantly burned to a crisp at 3000 times the speed of sound. Needless to say, more than a few parents were left with some difficult explanations for their heartbroken children. But we're grown-ups, right? We can *take* it! So just how fast *would* Santa have to go? Within an order of magnitude at least, was the vicar right? (More than one of these may apply.)

A) We'll never know. He's currently ministering to a parish of Yanomamo Indians in the Amazon rainforests of Brazil.
B) Perhaps he had the best of intentions, but he probably flunked math in high school. Even allowing for a one ten-thousandth of a second stopover in each home would still require travel at a rate in excess of the speed of light (let alone the speed of sound).
C) The vicar was right, at least within an order of magnitude.
D) Who cares? Just have a Merry Christmas!

Subject: Retractable *WHAT*?!

Question 19: True or false: There have also been retractable hull aircraft.

Subject: Unidentified Flying Hubcaps

Question 20: True or false: The United States military had in its possession, at one time, an actual flying saucer.

Subject: Talk About Inverted!

Question 21: True or False: An airplane can be made to land and take off while inverted.

Subject: Strength In Numbers

Question 22: Just as the greater relative proportion of one substance introduced into another can improve the effectiveness of a mixture or a solution in physical chemistry, the increasing numbers of a smaller population intermingled with another more predominant one can serve to further justify the agenda of the former (whether that be equal rights, or simply acceptance). When it comes to the pilot population then, at what point in time during the last 100 years has the relative number of pilots been the greatest, with respect to the population at large?

A) just before World War II
B) just after World War II
C) just after the Vietnam war
D) 1980

Subject: Pride Goeth Before A Fall

Question 23: Two pilots walk into the waiting room of an Aviation Medical Examiner to renew the medical certificates that they carry along with their private pilot certificates. One believes himself to be in excellent health, and although he doesn't need it, decides to "upgrade" to a first class medical. The other is content with keeping his third class medical. The "third class" pilot passes his exam with flying colors. The AME finds a minor eye problem however, in the case of the first guy. What happens?

A) The AME must re-examine the applicant according to Third Class criteria only. There is no "one level" demotion.
B) The AME must advise the applicant that he has to reschedule for another exam, and the applicant doesn't receive any medical at all.
C) The AME continues the exam, but must not inform the applicant until it is over as to the result, at which point the applicant has the option of receiving the next lower class of medical for which he qualifies, or coming back for another try.
D) The AME can put the exam on hold, pending further work by a specialist, for a reasonable period of time (say, two weeks). Then, if the problem proves not to be a show-stopper, he can get either his desired class of medical, or one rung (or two) down.

Subject: Off the Grid

Question 24: True or False: Airway intersections are always five characters, and always consist of five letters.

Subject: No, er...Kidding

Question 25: The term "voidance apron" means

A) a date and time after the expiration of a lease, after which a controlling municipality may appeal for rezoning of an airport property for other public purposes...such as...a park
B) a designated non-movement area at an airport where lavatory service trucks may discharge waste into the local sewer system
C) a polite term for an airline solution to bowel incontinence, also colloquially known as a "s**t wrap"
D) the correct term for the surrounding container of an oil reservoir when drive or transmission gears are splash-lubricated

Subject: “At the tone, the time is...”

Question 26: At Fort Collins, Colorado, and in Hawaii, radio stations fed from atomic clocks with an accuracy of 1 sec/million years, transmit continuous time signals throughout the world on 2.5, 5.0, 10.0 and 15.0 MHz. Since Fort Collins (Station code WWV) and Hawaii (Station code WWVH) transmit on the same frequencies, time statements are made

A) in English from Colorado and in French from Hawaii
B) with a human voice from Colorado and a computer generated voice in Hawaii
C) by a male voice from Colorado and a female voice from Hawaii.
D) on even minutes from Colorado, and odd minutes from Hawaii

Subject: Different Meanings, II

Question 27: What do the letters "Tango Uniform" signify when spoken?

A) A highjacking is in progress.
B) A flight instrument or control has ceased to function
C) nothing
D) a request for retransmission of your last message

Subject: Going Incognito

Question 28: The FAA requires many aircraft to display N-numbers that are a foot high. Which of the following are exceptions? (More than one may apply.)

A) Antique aircraft (a U.S. registered aircraft built more than 30 years ago, or an experimental category aircraft used either for exhibition purposes or amateur operation and having the same external configuration as one built at least that long ago) may have N-numbers only *two inches* high.
B) Experimental aircraft having a maximum cruising speed of 180 knots or less can use N-numbers that are three inches high (but they have to be on either both sides of the fuselage or both sides of the vertical tail surface).

C) all airships, gliders, or balloons, but again, it must be on opposite sides (in the case of balloons, at points near the maximum circumference)
D) An aircraft can get away with having <u>no</u> N-number displayed during an air show or the filming of a motion picture (or for practice and test flights, as well as en route to and from a base of operations).
E) If an airplane was manufactured before November 1, 1981 and originally displayed two-inch marks, they may be left that way until it is repainted or the markings are otherwise repainted, restored, or changed.
F) none of the above

Other Things You'd *Never* Think About: Answers

Answer to Question #1: The answer is D. I mean, C. (I was the lead engineer on GPS IIF crosslink, so I *know*!) WAAS of course (Wide Area Augmentation System) *also* increases accuracy, but unlike crosslink, this one came through.

Answer to Question #2: The answer is C. (I couldn't have made up a name like Reuchlin if I'd thought about it all day.) I first discovered this delectable item several years ago, when I came across <u>The Bishop's Boys</u>. My son had a science project and he had to interview Tom Crouch and Peter Jakab, NASM historians and Wright Brothers experts. Along the way, I discovered that wonderful book, written by Tom Crouch. (Reuchlin was pronounced ROOSH-lin, incidentally.)

Answer to Question #3: The correct answer is actually, and quite nobly, choice C.

Answer to Question #4: The answer is B. Decolletage is, um, well . . . look it up.

Answer to Question #5: The answer is C. Incredible but true. In China, even an automobile driver's license is about as unusual--and as hard to get—as a pilots license is, in the US. It was only recently in fact, in November of 2003, that the first light aircraft was purchased for private use on the Chinese mainland. (It was a Robinson R44 helicopter.) The market for light aircraft in China is projected to grow rapidly over the next few years (though there are still only about two or three hundred privately-owned aircraft in China)

Answer to Question #6: The answer is C. Advisory Circular 99-1C, Security Control of Air Traffic, describes it, as does FAR 99 somewhat. There is also a less-extreme area or regional operational state called "ESCAT" (Emergency Security Control of Air Traffic Procedures), which is what we saw happening on September 11, where the military mostly delegates air traffic control facilities to broadcast instructions (as in "everybody land, now!") You don't even want to <u>think</u> about a real SCATANA (or the most ominous one, OPLAN).

Answer to Question #7: D. Think about the ideal gas law $PV = nRT$ (where n is the number of moles, R is the gas constant, and P, V, and T are pressure, volume, and temperature, respectively). If volume is held constant, then when you raise the pressure, the temperature will rise accordingly. If heat were not extracted from the compressed air, temperatures could reach as high as 130 degrees F.

Answer to Question #8: C. Transponders are not just for airplanes! Ships at sea are often assigned codes in the range of 5200. (Actually, that EZPass toll thingy you may have in your car is also a transponder of sorts.)

Answer to Question #9: B. No, you didn't suddenly forget the two basic types of drag you learned about in private pilot ground school, namely parasite drag (form, skin friction, and interference) as well as induced drag from the development of lift. Nor is this a coy reference to things scatological. Excrescence drag is just a fancy word for skin friction drag, not really a slightly finer shade of meaning. It's simpler name could be "roughness" drag. It is expressed empirically in terms of the extra weight that would require the same amount of extra fuel to be burned, in the absence of such aeronautical warts and carbuncles. This esoterica matters mostly to wind tunnel weenies. A surface is considered hydraulically smooth when its "excrescences" lie within the boundary layer.

Answer to Question #10: C. It's available free, online, and it will even tell you the latitude and longitude for "your" intersection. If your first, middle, or last name has five letters--which isn't at all uncommon--chances are good that you already "have" an intersection.

Answer to Question #11: D (pretty boring, huh?)

Answer to Question #12: D. There would be a few limitations. Assuming that you had a calibrated pitot tube and a properly positioned static port installed, your airspeed indicator wouldn't register much below 45 mph, your altimeter would be useless unless you either always set it to 29.92 or always had the current atmospheric pressure, and you would have to buy a pre-swung compass to go with your DG. Furthermore, your attitude indicator would probably mostly play dead unless you lived in San Francisco, as would your VSI unless you drove somewhere like West Virginia. Your turn coordinator would almost always be showing a slip or skid, and you would almost always peg its poor wings during even the most modest turns around any suburban corner. You wouldn't need a transformer for the additional voltage, because a car battery and an aircraft battery are both 12 volts, and both of their alternators put out the same voltage: 13.5 to 14.5 V; they're both 14 V systems. Also, the current draw is nominal. Finally, there's more: the gyroscopic instruments would work, but only if they were either all-electric, or you added a vacuum system to power a conventional pair of attitude and heading indicators. (Don't expect much help on the Form 337 from your local FSDO, though...)

Answer to Question #13: C *and* E. The average 172, minus the seats, instrumentation, tires, plexiglass, and all other non-metallic parts, as well as all parts that are *not* made of aluminum, contains approximately 740 pounds of this wonderful metal. This information, using a 1976 model, comes straight from Cessna's Technical Services group in Wichita. This figure is incidentally also fairly close to the amount contained in a late model Skyhawk. An average empty soda can (say, a can of caffeine-free Coke) weighs about 13.5 grams (with the pull-tab still attached). That would come out to 33.6 cans per pound. That, multiplied by 740 pounds, is 24,864 or around 25,000 cans. As for choice E, there's a bit of an explanation there. Aluminum is usually alloyed with other elements such as copper, magnesium, manganese, or zinc, which is indicated by the first of four digits in the alloy designator series. Soda cans are made with alloy 3003 for example, and contain around 1% manganese. But the aluminum alloys used in aircraft construction are the 2xxx series (such as 2014) which uses copper, or the 7xxx series (such as 7075) which uses zinc...more than you ever wanted to know, I'm sure...

Answer to Question #14: A. No, it's considerably more than a school, or a mansion (though not quite as much as the B-2, which is over one *billion* dallars a copy). The going price for a B777 is about 135 million dollars. Something approaching half of that is for the computer systems and flight instrumentation alone. The number may seem staggering, but costs such as this are not that much more, in an order-of-magnitude sense at least, than for most large commercial transport category airplanes, and even such huge investments are more than amortized over their expected lifetimes. (For example, if you assumed a 300 seat configuration, that would come to $450,000 per seat. Assuming a per-seat cost of $200 per flight, that would require 2,250 flights, and assuming just two flights per day, the airplane could pay for itself in just three years.)

Answer to Question #15: C. Takeoff noise levels of 136 dB were measured, with perceived noise levels 10 dB higher than this in the 125 to 250 Hz frequency range (from NASA Technical Memorandum X-1565, April 1968). The approximate noise level of the B-1B at similar distances is 120 dB. The takeoff noise of Concorde was actually not too bad, relatively speaking (perhaps 115 dB), and was in fact a bit quieter than a 747. To give you some relative reference, a chain saw at normal operating distances puts out about 110 dB, and a very loud rock band perhaps 120 dB. (Keep in mind that the scale is logarithmic, and not linear.) Anything over 130 dB risks an actual excitation of structural component resonances. (Can you say "Jericho"?)

Answer to Question #16: C, and if you stretch things a bit, B. Although there are very tiny jet engines, such as the Jet-X (choice A), which produces six ounces of thrust; the JF-35 Nano-Bee, which has about 2.5 Kg of thrust; the AMT Olympus (choice B), which yields about 43 lb thrust; or the Cobra turbojet which at 6.2 pounds, produces 30 pounds of thrust, these are all almost always used by model aircraft enthususiasts. (And for those purists out there, yes, I know that pounds and kilograms are not, strictly speaking, the proper units of thrust, and it should be lbf or Newtons.) The TRS-18 used by the 12 foot long Bud Light microjet, which produces 360 lb of thrust and propels it to speeds of about 300 knots, is about a foot wide and two feet long, and weighs 75 pounds. In comparison, the next smallest jet engine that is (or hopefully, will be) used on a passenger aircraft is the Pratt & Whitney PW610F turbofan for the Eclipse 500. Uninstalled, it weighs between 200 and 250 pounds, and produces about 900 lbf of static thrust. The lightest commercially produced "real" jet engines may be the F107/F112s used in ALCM and Tomahawk missiles; they weigh about 146 pounds. If there were a smaller jet engine on say, an ultralight, it would need to produce somewhere around 50 lbf thrust. Actually, a French pilot has installed two AMT Olympus turbines in his Cri-Cri (Cricket) "ultralight". The original design has an empty weight of about 180 pounds, so by FAA CFR Title 14 Part 103 definition at least according to weight, it would be an ultralight (although it's actually not, because it exceeds the maximum 55-knot speed by quite a bit). Each engine has an installed total system weight of about seven pounds, and provides a thrust of about 43 pounds. The pilot, Nicolas Charmont, has acheived a top speed of about 150 mph (and he gets 100 mph with one engine).

Answer to Question #17: E, although possibly D. There can be no exact answer to this question (other than that it is definitely a long shot, well beyond the "one in ten million" range). From the standpoint of the probability or acceptability of operational errors or mechanical malfunctions in designing or operating any aviation device or system, the FAA uses terms to define the odds of something going wrong. One of these is "extremely improbable". Although not universally applied, this term, as well as the others, are defined in Advisory Circular 120-29.

I asked Bruce Landsberg, Director of the AOPA Air Safety Foundation, if anyone had ever experienced such a failure, and he didn't think that it had ever actually happened. As he pointed out to me, almost all of the spatial disorientation accidents are due to continued VFR into IMC; on average, there are only three accidents per year because of spatial disorientation that was caused by loss of vacuum or actual instrument failure. (During a recent 10-year period, the actual number of general aviation "disorientation related" accidents was exactly 271, but in fact, *only 17 of those were due to a vacuum system or pump failure, just six were because of an instrument failure, and a very unlucky but very small number, just two, had both*. The remaining 246, and the vast majority, were simply pilot proficiency or judgment screw-ups.) So don't get too worried on this one. First of all, most of your flying is likely not to be in IMC. Secondly, most electrical failures don't happen all at once. (The trick is to keep the ammeter in your scan, or at least to look at it every couple of minutes, while in IMC. If the alternator does take a vacation, you do have a battery, which confers a precious few minutes to try getting to visual conditions.) You may also have a hand-held backup radio. And you might have a standby vacuum (no, not the one for the carpet). Also, you might periodically practice partial panel flying...

So here's the big picture: Each year, general aviation flies somewhere on the order of 30 million hours. For the last 30 years, it's been between about 24 and 43 million, with the high in the late 1970s, and a low in the mid-90s. Considering just the last 30 years, that would be, very roughly, a billion hours flown. And it's probably never happened (or at least been reported). Heck, although it depends on where you live and what you do, in your entire lifetime the odds of the average American being struck by lightning are about a third of a million times greater. (According to the National Weather Service, it's one in 3000.)

Answer to Question #18: B and D. No, we're not going to deal with actual populations, shortest-path algorithms, graph theory, or any of that stuff. We're only interested in seeing if he was even in the right "power of ten" ballpark (i.e., somewhere between ten times slower and ten times faster than what he'd said). If you assume the shortest path between a few hundred million Christian homes can be represented by a simple-minded rotation of great circle arcs about both poles, separated by, say, 100 feet at 30 degrees North, and making the assumption that if he did equator-ward course reversals upon reaching 60 degrees North and 60 degrees South, he wouldn't miss too many chimneys...*and* if you even factored in an additional ten degrees northward travel for Alaska, Sweden, Norway, and Finland (sorry, Siberia) that would be, let's see...

The earth's equatorial radius is about 3963 statute miles, it's equatorial circumference is about 24902 miles, it's circumpolar circumference is about 41 miles less or roughly 24860 miles. If you used 30 degrees of latitude as a reference, 3963 times the cosine of 30 degrees gives a distance from the earth's axis at that latitude of…3432 statute miles. That, times two pi, gives a circle at 30 degrees latitude of 21564 miles. Multiplying by 5280 feet, then dividing by 100 feet equals 1,138,591 successive longitudinal course lines, between 60N and 60S, around the earth. Let's take the liberty of being charitable to Santa, and chop that in half, assuming he can skip those bands where there are only vast stretches of ocean (sorry, Tahiti) and round it down to just 569,000. If you then estimate the distance along 120 latitudinal degrees of arc (along each of those course lines) of roughly 8290 miles (using about one third of the circumference), that would give 4,717,010,000 or about four billion, 717 million miles. Then if you added in a bit extra (about half of one percent) for Alaska and northern Europe, it would wind up being about four and three quarter billion miles. Let's say we allowed Santa & Co. an eight hour shift. (That's not against union rules, right?) Actually though, being such a consummate globe-trotter, Santa has a big advantage. He can "follow the night", so instead of only eight hours aloft, as far as the *whole world* is concerned, Christmas eve for him would actually be 32 hours! (Why? Well, the rotating earth could keep 24 of them at bay until eight *more* were used up in that lightning-fast feat of gift-giving.) Okay, so covering that distance in 32 hours means a speed of roughly 148,440,000 miles an hour, and dividing that by the speed of sound (at sea level of course) of about 761 mph would give about Mach 195,000. (And that's *without stopping anywhere.*) That's also--ahem--about 22 percent of the speed of light, or for physics weenies, 0.22c. And if you're wondering about relativity, it would actually work *against* him, as time would go by on earth some two and a half percent *faster* (from Santa's perspective, that is). Yes, from a non-relativistic standpoint, you'd think that he could get more work done. By the time he'd reached 71% of the speed of light, he could make the whole trip in about 10 hours, even though the "time dilation" penalty would only be about 40%. It gets worse from there, though. At 0.98c he might do it four and a half times faster than he could at 0.22c, but thanks to special relativity, time would be going by on earth *five* times faster! (and at 99% of the speed of light, seven times faster...) So, yup, even if the assumptions behind my calculations are off by a factor of *ten*...he's probably still wrong...unless of course, he just meant Jolly Olde England. Then...well...he would be just about right.

By the way, if you'd like to track Santa's progress (although I have no idea how NORAD can actually *do* that, given his extreme speed), you can follow along at their "NORAD Tracks Santa" web site, noradsanta.org.

Answer to Question #19: True...in a manner of speaking, of course. (One may justifiably wonder about just what other dimension there might be, into which a hull itself might possibly retract!) There are retractable landing gear, someone tried retractable (telescoping) wings, some motor-gliders have retractable propellers, so why not try a retractable hull? Britain's Blackburn Aircraft Company built an aircraft, the B20 (a seaplane) with such a so-called retractable hull. Cumbersome "stepped" hulls, necessary to allow the hull to overcome surface tension during takeoff before it could climb up "on the step" (as well as keeping the prop above the spray), or clumsy flat-sided hulls were inefficient once airborne. Their top speeds didn't get much over 200 mph. The twin-engine B-20 had a watertight hull section attached to the fuselage by folding struts. Upon looking at this picture, one might conceivably be tempted to think this was nothing more than one big retractable pontoon and an overblown topological non-sequitur, but since it was in fact a section of the lower hull, it qualifies (sorta). Once airborne, the pilot retracted this extended hull segment snugly against the bottom of the fuselage. Speeds in excess of 300 mph were expected, but sadly, the prototype crashed in 1940, and then the project was abandoned in favor of higher priority war efforts.

photo from "The Blackburn Story 1909-1959" originally published by the Blackburn Aircraft Company

Answer to Question #20: True. Actually, they had two. Notice though, that I didn't say anything about how well they flew (or where they came from). During the middle of the twentieth century, the US military oversaw the development of an extraordinary variety of test aircraft, testing many variations of airframes and means of propulsion. In the early 1950s, a Canadian firm named Avro Aircraft—the parent company was A.V. Roe—began developing a unique, supersonic VTOL fighter-bomber aircraft intended to cruise at low altitudes on a cushion of air and accelerate to high speeds at high altitudes. In 1958, the U. S. Army and Air Force took control of the project, naming the vehicle "Avrocar" and designated it the VZ-9AV ('VZ,' experimental vertical, ninth concept proposal, and Avro). The Army wanted a rugged subsonic, all-terrain vehicle: in short, a flying jeep. The Air Force wanted something that could hover beneath enemy radar and then rocket into the stratosphere at supersonic speeds. The entire aircraft was a circular wing not quite 20 feet wide, three feet thick, and shaped exactly like a flying saucer. The aircraft was powered by three gas-turbine engines which drove a five-foot "turborotor" mounted in the center of the vehicle, with thrust directed through nozzles and peripheral jets for lift and control forces.

A scale model of the aircraft sent to Wright Field revealed a serious problem: at altitudes of more than a few feet the cushion of air supporting the aircraft became unstable, causing uncontrollable pitch oscillations, coupled with heavy roll. (Avro called it "hubcapping".) The problem could be solved but it would probably never fly supersonic. Despite this, the US hoped that at least the Army's requirements could be met. In the fall of 1959 the first completed prototype rolled out onto Avro's taxiway near Toronto. The Avrocar first flew with a pilot on Dec. 5, 1959. The VZ-9 reached a maximum speed of 35 mph, and its service ceiling turned out to be all of about six feet; all attempts to eliminate hubcapping failed. A second prototype was shipped to NASA Ames in California, for wind tunnel tests. Exploring the other end of the flight envelope, they discovered the Avro had insufficient control for high-speed flight and was aerodynamically unstable. Adding a conventional empennage did not improve things. The technical problems seemed insurmountable and the U. S. Air Force terminated the program in December 1961 due to limited flight success, mechanical problems, and possible disenchantment over the flying saucer design. The Avrocar was the last aviation program of Avro Canada, and the parent company ceased to exist by May 1962. Two examples of the saucer ship still exist, though; one is at the US Army Transportation Museum at Fort Eustis, Virginia, the other was at the Smithsonian's Garber restoration facility, and is scheduled for display at the new Udvar-Hazy Center in Chantilly, Virginia.

flying saucer type VZ-9AV "Avrocar"

Answer to Question #21: Okay, I'll bet you're wondering how I'm going to weasel out of *this* one. Well, this picture ought to pretty much give it away. Nicknamed "Double Take", this Pitts S-2B was owned by one Craig Bryant Hosking of California, who is a Hollywood aerobatic and stunt pilot (and a recipient of the Art Scholl Memorial Award). His name is in the credits for *Space Cowboys*, *Pearl Harbor*, and *The Kid*, among others.) Craig first demonstrated inverted landings in the mid-1980s.

He selected the Pitts S-2B for its powerful engine, its ability to maneuver through most aerobatic routines, its durable airframe, and outstanding performance. Modifications were planned and drawn up with the help of a fellow aerobatic pilot, aircraft builder, and welder. Aside from having to attain a new perspective on landing an airplane, an obvious requirement to attach a second set of landing gear on the top wing, a special retention system that allows him to be lowered from the cockpit while inverted on the ground, others included ensuring a second brake system, and keeping the cylinders free from oil when the plane was parked inverted. It has since retired from airshow flying and has been donated to a museum.

"Double Take", Craig Bryant Hosking's modified Pitts S-2B

Answer to Question #22: Based on historical data for the US *civilian* pilot population from AOPA, the peak in the total number of American pilots wasn't before or after WWII, or the Vietnam war, but a few years afterwards in the late 1970s (actually 1980, at over 827,000). That is a huge number compared to the present, especially since the US population in 1980 was only about four-fifths of what it is now. The US pilot population was about one third greater back then, which means that the pilot-to-groundling ratio has since dropped by about 40%. (Four fifths times three quarters equals exactly six tenths.) That was the year when the percentage of the general population who were pilots was greatest: not quite four tenths of one percent. Here are a couple of graphs to illustrate this. (The known total US population data points are only once every ten years, due to the census being done only once each decade, and AOPA's pilot population datum for 1950 is suspect, so I didn't use it. Also, due to changes in FAA procedures, and other record-keeping related glitches, some data points may be off somewhat, but it's the best we have.) So, bottom line, the correct answer is D.

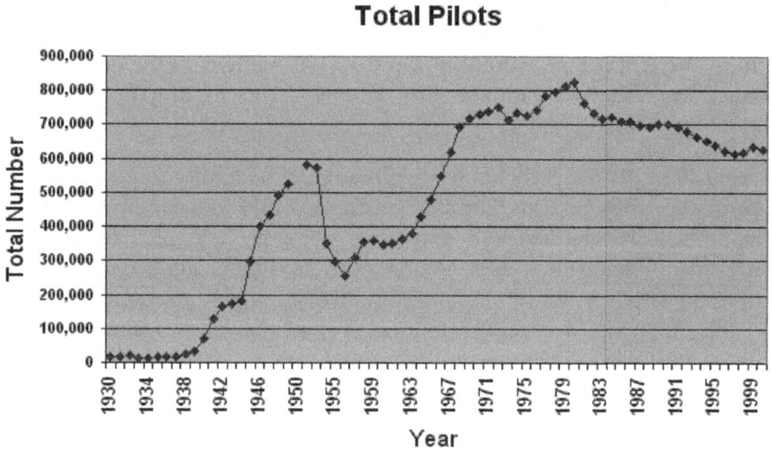

The first rapid rise in the number of pilots was due primarily to the Civilian Pilot Training Program, which arose from verbiage in the Civil Aeronautics Act of 1938 and which was unveiled by President Roosevelt late that year. It started shortly after that, early in 1939. Uncle Sam paid for 72 hours of ground school and between 35 and 50 hours of flight instruction. Although the military was skeptical and even among the civilian population, support was divided along party lines, the invasion of Poland in September of that year made the program's value and importance quite clear, and the invasion of Pearl Harbor removed any further doubt. Once at war, the CPTP became the War Training Service, and unlike its predecessor, all graduates entered military service. (Perhaps a few years later, the reality of the workforce and family obligations accounts for the sudden drop-off.) There was a second upsurge throughout the 60s and this may be Vietnam era related (pilots exiting the military or using their GI benefits), or simply a demographic result of a healthy economy and increased discretionary income. Most likely it is simply the "second generation" phenomenon.

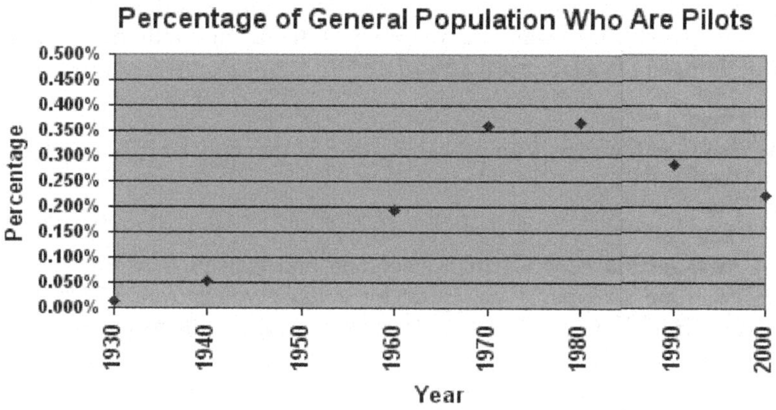

graphs created by the author

Answer to Question #23: You won't find this in Part 61, 91, or anywhere else in the usual FARs. According to Dick Hiner, the former vice president of training at the AOPA Air Safety Foundation, when you apply for a medical certificate, you apply for the specific class on the application. If you don't meet the requirements for the class for which you applied, the application must be rejected, and you walk out the door without any medical at all. The only thing that one can do is fill out another application for a lower class. However, you must reschedule with the AME. Not only that, but in a subsequent application, you are supposed to indicate that you have failed a previous medical exam and as you can guess, this opens up a can of worms. The best guidance then is to go for it, but only if you have a valid need or reason. So the answer, according to this, is B. (I did ask a local senior Aviation Medical Examiner about this situation and he indicated that there *could* be some leniency, but don't count on it.)

Answer to Question #24: Those familiar intersections, as specified in FAA publication 7350.7G, LOCATION IDENTIFIERS, do indeed consist of five characters, all alphabetic. However, there are also *other* types of intersections. Did you know for example that there are intersections which are also five characters long, but which consist entirely of numbers? These are used, not by pilots (although they *can* be) but by the Flight Data Center, whenever airways intersect other airways at otherwise "non-intersection" points. Any time you see Victor airways crossing, with no named intersection, it actually does have a name; it's just not charted. There are also "High Altitude Redesign" alphanumeric waypoints (initially every two degrees of longitude and 30 minutes of latitude) for FL 390 and above, which are anticipated to be placed at every degree of longitude and 10 minutes of latitude. They will have names starting with the letter "K", then another letter (going roughly from A to Z) that will represent the ARTCC containing the waypoint, followed by two digits from 00 to 90 (for latitude) and finally another letter (for a longitude increment). So, no, they're not just alphabetic. Most of the ones we're familiar with are, but they can also be entirely numeric, or a mix. The answer is: "false". (You can read more about these high-altitude waypoints in Advisory Circular 90-99, dated 9/22/2003.)

Answer to Question #25: You probably guessed this one. (It's the most indecent and indecorous: It's C)

Answer to Question #26: C. Since Colorado and Hawaii transmit on the same frequencies, time statements are made by a male voice from Colorado and a female voice from Hawaii. (At least, that's one way to tell them apart.)

Answer to Question #27: B. Although this is definitely not approved phraseology, and is hardly politically correct, this radio-phonetic pronunciation for the letters T and U has been used colloquially to mean "tits up". In normal usage it meant flat-on-your-back dead. Defunct. Moribund. Inert. Inoperative. When your radio quits, or your engine stops, or your lights go out, they're Tango Uniform.

Answer to Question #28: With the exception of the last choice (F), all of them. See Advisory Circular 45-2B, dated 7/16/03. Also, 14 CFR part 45 describes the rules for displaying registration markings and identification plates, and Advisory Circulars 43-3 and 43-17 cover the installation, removal, or change of aircraft identification data.

About the Author

Jeff holds physical science and engineering degrees and his vocational interests over the last 30 years have been in the telecommunications industry. His primary avocation during the last 25 has been aviation, and he is a commercial pilot with ASEL, instrument, helicopter, and glider ratings. He has flown as a mission pilot for the Civil Air Patrol and Angel Flight. His outlook on life can best be described as one of philosophical whimsy, with an engineer's affinity for details. He's written hundreds of aviation articles for iPilot.com, as well as several dozen for AOPA. He lives in New Market, Maryland.

Made in United States
North Haven, CT
05 November 2026

21766152R00174